T0281099

Cambridge Elements ≡

Elements of Paleontology
edited by
Colin D. Sumrall
University of Tennessee

PHYLOGENETIC COMPARATIVE METHODS: A USER'S GUIDE FOR PALEONTOLOGISTS

Laura C. Soul
The Natural History Museum, London and National Museum of Natural History, Smithsonian Institution

David F. Wright
American Museum of Natural History and National Museum of Natural History, Smithsonian Institution

Paleontological
S O C I E T Y

CAMBRIDGE
UNIVERSITY PRESS

CAMBRIDGE
UNIVERSITY PRESS

University Printing House, Cambridge CB2 8BS, United Kingdom

One Liberty Plaza, 20th Floor, New York, NY 10006, USA

477 Williamstown Road, Port Melbourne, VIC 3207, Australia

314–321, 3rd Floor, Plot 3, Splendor Forum, Jasola District Centre,
New Delhi – 110025, India

79 Anson Road, #06–04/06, Singapore 079906

Cambridge University Press is part of the University of Cambridge.

It furthers the University's mission by disseminating knowledge in the pursuit of
education, learning, and research at the highest international levels of excellence.

www.cambridge.org
Information on this title: www.cambridge.org/9781108794688
DOI: 10.1017/9781108894142

© Laura C. Soul and David F. Wright 2021

First published 2021

A catalogue record for this publication is available from the British Library.

ISBN 978-1-108-79468-8 Paperback
ISSN 2517-780X (online)
ISSN 2517-7796 (print)

Additional resources for this publication at cambridge.org/soulwright

Phylogenetic Comparative Methods:
A User's Guide for Paleontologists

Elements of Paleontology

DOI: 10.1017/9781108894142
First published online: April 2021

Laura C. Soul
The Natural History Museum, London and National Museum of Natural History, Smithsonian Institution

David F. Wright
American Museum of Natural History and National Museum of Natural History, Smithsonian Institution

Author for correspondence: Laura C. Soul, laura.soul@nhm.ac.uk

Abstract: Recent advances in statistical approaches called phylogenetic comparative methods (PCMs) have provided paleontologists with a powerful set of analytical tools for investigating evolutionary tempo and mode in fossil lineages. However, attempts to integrate PCMs with fossil data often present workers with practical challenges or unfamiliar literature. This Element presents guides to the theory behind and the application of PCMs with fossil taxa. Based on an empirical dataset of Paleozoic crinoids, example analyses are presented to illustrate common applications of PCMs to fossil data, including investigating patterns of correlated trait evolution and macroevolutionary models of morphological change. The authors emphasize the importance of accounting for sources of uncertainty and discuss how to evaluate model fit and adequacy. Finally, the authors discuss several promising methods for modeling heterogeneous evolutionary dynamics with fossil phylogenies. Integrating phylogeny-based approaches with the fossil record provides a rigorous, quantitative perspective on understanding key patterns in the history of life.

Keywords: phylogenetic comparative methods, paleontology, macroevolution, R, fossil

ISBNs: 9781108794688 (PB), 9781108894142 (OC)
ISSNs: 2517-780X (online), 2517-7796 (print)

Contents

1 Introduction

A fundamental prediction of biological evolution is that a species will most commonly share many characteristics with lineages from which it has recently diverged, and fewer characteristics with lineages from which it diverged further in the past. This principle, which results from descent with modification, is one of the most basic in biology (Darwin 1859). A reconstruction of the relationships between species that is based on similarities and differences in characteristics – a phylogenetic hypothesis – can be used as a powerful tool to understand fundamental questions about the history of life (Nunn 2011; Baum and Smith 2013; Harmon 2019).

Phylogenetic comparative methods (PCMs) can be broadly defined as statistical approaches that incorporate information about the shared evolutionary history of taxa (i.e. their nonindependence) to identify macroevolutionary patterns or test hypotheses about how those patterns relate to macroevolutionary drivers, such as climate or biotic interactions. Until recently, there have been largely separate analytical frameworks for phylogenetic inference (inferring relationships between taxonomic units using morphological or molecular character data) and PCMs (testing hypotheses about evolution while treating the relationships as known). In this Element we only discuss the latter, but other recent literature focuses on a unified methodological framework that integrates the two (Warnock and Wright 2020; Wright et al. 2020). Many other types of information, such as biogeographic data (Matzke and Wright 2016; Landis 2017), can be used in this modeling framework, and as new models are developed the scope of questions that can be addressed will increase.

There are many reasons (perhaps particularly for paleobiologists) why it might not be feasible, or of interest, to use this kind of unified framework, where you must infer phylogenetic relationships as a means to answer other macroevolutionary questions. Perhaps you have previously estimated a phylogeny but now want to use it to answer new questions, perhaps you are interested in combining several smaller phylogenies to generate a supertree, or perhaps your specific question is not yet answerable in a Bayesian process-based framework (e.g. Warnock and Wright 2020; Wright et al. 2020). The good news is that trees constructed in many different ways can be used in PCMs to make reliable inferences about trait diversification, provided that the tree is appropriately scaled to time using the stratigraphic record as an extra source of information (Bapst 2014a; Soul and Friedman 2015; Barido-Sottani, Tiel et al. 2020).

PCMs have been rapidly proliferating in the past five to ten years, and the kinds of questions they can be used to rigorously answer are now very diverse.

Here, we review and demonstrate some of the analytical approaches that can be applied when you already have a phylogeny in hand. Many of these PCMs are used to model trait change through time and the relationship between that trait change and other variables. Most do not model the underlying microevolutionary processes occurring in populations or the external drivers that generate the trait change. Instead, PCMs more commonly use stochastic models to investigate the long-term outcome of evolutionary change. Therefore, as is outlined in more detail later on, different underlying processes can generate similar patterns that can be equally well explained by the same model. Careful interpretation of the results of any analysis is imperative.

We begin by outlining some fundamental approaches that are conceptually important, and then move on to more complex macroevolutionary models. Multiple books have been written on this vast topic (see Nunn 2011, Garamszegi 2014, and Harmon 2019, each of which are not focused on the fossil record but paleontologists might find them useful nonetheless), but we hope to provide you with a clear, digestible explanation of the theory behind a variety of PCMs, along with enough information on how to apply them to get started in answering your own questions. A reader already well versed in these approaches will find a review of recently published methods, and suggestions for their implementation in a paleobiological context, in the last third of the Element. A variety of software has been used to implement PCMs, but the majority of those likely to be of interest to paleontologists are available in R (e.g. Bapst 2012; Lloyd 2016; Barido-Sottani et al. 2019). We therefore provide all of our reproducible examples in R. Package names and inline code examples are in `Courier New` font.

All data used in the examples in this Element are available on GitHub (https://github.com/daveyfwright/PCMsForPaleontologists), along with a script to load and format the data in the R programming environment so that they can be analyzed immediately, as well as the full R script and annotated script and outputs from the example analyses. Although the examples in this Element are intended to provide a guide for implementing comparative analyses in R, we encourage readers to also follow along directly using the scripts we provide on GitHub. The example code in the main Element assumes a basic working knowledge of R, including loading data, inspecting objects, reading help pages, and object assignment. The materials online do not make this assumption and provide a detailed step-by-step guide.

2 Getting Started: Data and Phylogeny

Functions in R that can be used to manipulate phylogeny or apply phylogenetic comparative approaches make use of trees that are in "`phylo`" format, originally implemented in the ape package (Paradis et al. 2004). In this section, we outline features of this format, as well as some important things to remember when preparing paleontological data for an analysis in R (see also Bapst 2014b). The dataset we use here is for fossil crinoids (Eucladida, Echinodermata), a morphologically diverse clade of marine invertebrates with a well-sampled fossil record. Our use of this dataset is primarily intended to demonstrate the different tree-based analytical tools that can be applied, rather than to glean specific inferences about crinoid macroevolution, and we caution readers that our results should be viewed in this light.

Key packages in R that contain implementations of PCMs that are commonly applied to fossil data are `ape` (standard format and processing for phylogenies in R – Paradis et al. 2004), `nlme` (fitting Gaussian models, e.g. least-squares regression – Pinheiro et al. 2019), `geiger` (a versatile package that performs and plots many PCMs – Harmon et al. 2008; Pennell et al. 2014), `phytools` (additional plotting and simulation functions – Revell 2012), and `OUwie` (heterogeneous macroevolutionary model fitting, e.g. Brownian motion or Early Burst – Beaulieu and O'Meara 2020). There are many others, so it is valuable to spend time exploring the different tools that might be best suited to answering a particular question. Once you have gained familiarity with the above key packages, a very extensive list of all the packages that can be used for phylogenetic approaches in R can be found on this website: https://cran.r -project.org/web/views/Phylogenetics.html.

2.1 Phylogeny

For most of the example analyses we use a single phylogeny of eucladid crinoids that has branch lengths that represent time in millions of years (Figure 1, called `tree` in our scripts). It is the maximum clade credibility tree (MCCT), inferred using a model of morphological character evolution combined with a process-based model of diversification that allows inference of ancestor–descendant relationships (this method is distinct from ancestral *state* reconstruction – see subsection 4.2; Wright 2017a). The MCCT is the tree in the posterior distribution with the largest product of clade frequencies (i.e. probabilities), which represents a point estimate of phylogeny in a Bayesian context. Our emphasis on a single tree is for illustrative purposes only; in reality, analyses should *always* be applied to a set of possible phylogenies to assess how robust results are to variation in tree topology and branch lengths (see Section 6). The set of possible phylogenies (often referred to as a tree set) could be a set of the most parsimonious trees, a random sample from a Bayesian posterior distribution, repeats of stochastic

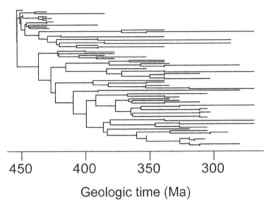

450   400   350   300

Geologic time (Ma)

Figure 1 Phylogenetic tree comprising eighty-two species belonging to the Eucladida (Crinoidea, Echinodermata). Tree topology and branch lengths correspond to the maximum clade credibility tree (MCCT) based on a Bayesian tip dating analysis of morphological character data presented in Wright (2017a). Branch lengths are scaled to absolute time in units of millions of years. The MCCT is called `tree` in our scripts and represents a point estimate of phylogeny. Where possible, we recommend testing macroevolutionary inferences across a distribution of trees rather than a point estimate, and our emphasis on the MCCT is for illustrative purposes only (see subsection 2.1 and Section 6) .

timescaling of a composite topology (Bapst 2014a), etc. The number of trees to include in the set depends on how variable results are across different trees; if results are highly variable, a larger set will be needed to characterize the possible outcomes of the analysis and how common they are. For example, 100, 500, or 1,000 are common set sizes in the published literature. The phylogenies used for analysis should have branch lengths that are scaled to time (best practice would be to use a birth-death-sampling approach [Stadler 2010; Gavryushkina et al. 2014; Heath et al. 2014; Wright 2017; Stadler et al. 2018], but if that is not possible, see Hedman 2010, Bapst 2014b, Halliday and Goswami 2016, or Lloyd 2016 for options and considerations when choosing an a posteriori timescaling approach).

A tree in `phylo` format has branches that are called "edges" and branching points that are called "nodes". For timescaled phylogenies, the branch (or edge) is a graphical representation of the amount of time since the lineage leading to one taxon diverged from its sister lineage, so the length of terminal branches does not represent the amount of time the actual species or genus at the tip was extant, only the time since divergence. Each node has a number assigned to it, beginning with the tips. This format is for both ultrametric (branches all end at the same time; usually trees of all living taxa) and nonultrametric trees (branches end at different times; usually trees that include extinct taxa).

A `phylo` tree in R has four components: (1) a matrix that identifies how the branches connect, by listing their start and end node numbers, (2) a vector of tip labels (user defined, usually taxon names), (3) a vector of branch lengths, and (4) an integer number that is the number of internal nodes. For trees that include only extinct taxa, the root age of the tree can be set using `tree$root.time <- X`, where X is the numerical age (usually reported in units Ma for paleontological datasets). A root age is required by some analyses, and facilitates good visualization when plotting a tree. If there is no root age assigned, most functions will assume the youngest tip ends at the present day; without it, the tree is in relative time, rather than absolute.

An important first step prior to analysis is to plot the tree (Revell et al. 2018). In our eucladid example some taxa were inferred to be ancestral to others in the tree (Figure 1; see subsection 4.2 for clarification of what this means). In R, a `phylo` object displays these sampled ancestors as sister to their descendants, but with zero-length branches (i.e. no inferred change between the node and tip for the ancestral taxon). For many PCMs, zero-length branches are mathematically intractable (the reasons relate to division-by-zero issues). Adding a very short length to each zero-length branch resolves this problem. If you are concerned about the possibility of this introducing a bias in your own analysis, you could also drop these tips from the tree and compare the results using each tree. Note that in the Eucladida dataset there are many inferred sampled ancestors, so dropping these tips may represent high information loss. Whether or not this is the case for different datasets will depend on the group under investigation. Node labels are required by some PCM packages (e.g. `OUwie`), so it is best practice to assign them using a vector the same length as the number of nodes (as long as you check that the R function you are using doesn't use them for anything you aren't expecting; for example, `OUwie` uses them to define ancestral macroevolutionary regimes – see subsection 9.1). This can be done using `tree$node.label <- rep(1, Nnode(tree))`, which gives all nodes the label "1".

A very useful basic function to inspect and manipulate phylogenies in R is `vcv`. When applied to a timescaled tree this function outputs the phylogenetic variance-covariance (VCV) matrix of that tree. The VCV matrix is an intuitive numeric representation of the tree, and is used in the inner workings of many R functions implementing PCMs. Models of continuous trait change (like Brownian motion; see Section 4 onwards) differ from one another with respect to how the variance and covariance of the trait are expected to change through time, and the VCV matrix is the basis for the statistical expectation under different models. Elements on the diagonal of the matrix give the variance. When a tree has branch lengths scaled to time (as paleontological trees usually do), these values give the root-to-tip distance for each taxon on the tree (i.e. duration). The functions `max(diag(vcv(tree)))` will output the maximum root-to-tip distance (i.e. the duration of the whole tree). The off-diagonal elements give the covariance, which is the amount of shared variance between pairs of taxa (i.e. the length of their shared evolutionary history). Figure 2 shows a small example tree and its associated VCV matrix.

2.2 Trait Data

In our example analyses we use a variety of PCMs to explore patterns in two continuous traits measured for each species in the Eucladida tree. These are `Shape` (calyx shape; the natural log of the length/width ratio of the calyx) and

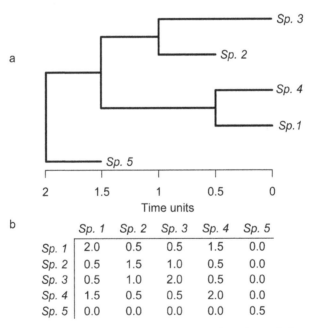

Figure 2 (a) Hypothetical phylogeny with branch lengths scaled to absolute time. (b) Variance-covariance matrix that describes the tree. The values along the diagonal of the matrix represent the amount of time from the root to each tip in the tree, which corresponds to their expected variance in a Brownian motion model. In contrast, the off-diagonal values represent the shared amount of evolutionary history for pairs of tips, which corresponds to their evolutionary covariance. Note that the diagonal values in the matrix represent terminal branches whereas the off-diagonal values represent internal branches.

Density (filtration fan density; the natural log of the approximate number of proximal feeding appendages an individual of the species has). We also briefly demonstrate analysis of a discrete trait Complexity (calyx complexity; the number of plates interrupting the posterior interray). We store all the traits in an object called alltraits for ease of use in R. Some analyses implemented in R require that the tree and data have exactly the same taxa; for these we use a tree (called prunedTree in code examples) in which taxa not present in the trait data have been removed from the tree. We use an estimated standard error of 0.03 for Shape and 0.12 for Density, based on an average across species for which there is more than one specimen.

3 Phylogenetic Nonindependence

Felsenstein (1985) was the first to outline the problem of the nonindependence of species trait data, and an algorithm to account for that problem, which he called phylogenetic independent contrasts (PIC). The most common evolutionary question that this problem (and proposed solution) is applied to involves the relationship between two traits. The extreme case of the problem of nonindependence of species data is shown in the original paper (Felsenstein 1985, figure 1), and a related example is provided in Figure 3 based on Nunn and Barton (2001). An early divergence within a clade leads to two groups of taxa that have quite different values for morphological traits between those two groups, and more similar trait values within each group, for both of the traits under investigation. When a linear regression model or correlation test is used, this will result in a strong relationship being inferred, but effectively a line is being fit to two points – the two groups within the clade, and no such strong relationship in fact exists. The early split means that species in the two parts of the tree have been evolving separately from one another for a long time, and so have had a long time to accumulate differences between them; species in the same part of the tree are more similar to each other because of their recent common ancestor and long shared evolutionary history. Regression analysis assumes that individual data points are statistically independent from one another; this assumption is violated by species data because of the shared evolutionary history.

PIC analysis is an intuitive approach to this problem. It takes the average trait value of sister clades and weights this by the amount of time since their most recent common ancestor (i.e. the amount of time they have been evolving separately). Most researchers now use phylogenetic generalized least squares (PGLS) instead of PIC because it is a more flexible likelihood-based model framework (PIC can be shown to be a special case of PGLS; see Garland and Ives 2000; Blomberg et al. 2012). After a nonphylogenetic ordinary least-

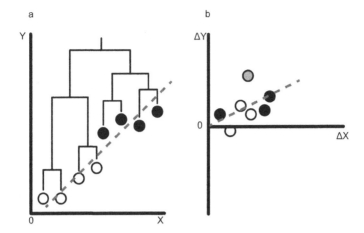

Figure 3 How species nonindependence may influence trait correlations and how phylogenetic independent contrasts address the issue. (a) Raw values for two traits (X and Y) plotted alongside the underlying phylogenetic relationships among species. Two subclades are identified and labeled by open and filled circles, corresponding to species with smaller (open) vs. larger (filled) trait values. The dashed line represents a best-fit linear model using ordinary least-squares (OLS) regression. (b) Phylogenetic independent contrasts estimated as the standardized difference between species trait values for each internal node in the tree. Note that the y-intercept goes through the origin, and the difference in slope. The contrast with the largest absolute value (gray circle) represents the evolutionary shift in trait values between the two subclades. Adapted from Nunn and Barton (2001).

squares (OLS) regression is performed on species data, there will be a phylogenetic signal that dictates how far each datapoint is from the regression line (its residual). We say therefore that there is phylogenetic structure in the residuals of the regression. PGLS incorporates information from the phylogeny into the regression model to adjust the regression line so that the residuals are normally distributed, rather than structured, rendering the analysis statistically valid.

Fitting both an OLS and PGLS regression line can be done with the same R function, `gls`. Generalized least-squares regression is a standard approach to regression that can be used when the datapoints are not independent from one another. When this is the case, the residuals of a standard regression will have a structure that is caused by the relationship between the datapoints. Generalized least-squares regression is a very flexible framework that allows you to supply a correlation structure for the residuals of the regression that can be derived from any potential source of nonindependence. For example, here we are using phylogeny, so we supply the phylogenetic VCV matrix to define the expected structure in the residuals, but, in another example, because some traits are known to vary systematically across space (e.g. Wagner and Marcot 2010), a matrix of expected spatial covariance could also be used.

There has been repeated discussion in the literature about whether researchers should or should not "correct for phylogeny" in a particular analysis by using approaches like PIC (Felsenstein 1985; Harvey et al. 1995; Rohlf 2006; Westoby et al. 2016) and a concern about "overcorrecting." The original formulation and explanatory figure for PIC has led to this terminology, which in turn may have led to some misunderstandings. Although it is true that without considering phylogeny, a comparative analysis of two species variables might be statistically invalid, it might be helpful to think of PIC (and PGLS) as incorporating the additional useful information that phylogeny provides into the estimation of the relationships between traits, rather than as a correction.

In the context of analyses like regression or ANOVA that can be used to understand trait correlations and adaptation, if there is a phylogenetic signal in the *residuals* from a model fit, then the resulting relationship derived is statistically invalid because the assumption of independent datapoints has been violated. Phylogenetic signal in any of the individual traits under investigation does not necessarily mean the residuals of the regression will have a phylogenetic structure, or vice versa. Revell (2010) provides a thorough explanation of this issue. If in doubt, rather than using the function corBrownian shown in Example Analysis 1, you can use corPagel to define the correlation structure. This will jointly estimate lambda, which is a measure of the phylogenetic signal in the residuals. As lambda gets closer to 0, the estimated coefficients in PGLS will converge on those estimated using OLS. Think about what that means in terms of model assumptions between OLS and PGLS. Just because "nonphylogenetic" methods like OLS do not incorporate information about evolutionary relationships does not mean they do not make assumptions about evolutionary change. In fact, the biological assumptions underlying OLS for comparative analyses are mathematically and conceptually equivalent to using a phylogeny, and assuming that it is a star phylogeny (i.e. all branches simultaneously diverge from a single node). Thus, both OLS and PGLS are based on models. It is important to keep in mind that all models are wrong, but some are more wrong than others. We advocate for using model goodness-of-fit tests like the Akaike information criterion (AIC) and Bayesian information criterion (BIC), or, where possible, going further and investigating model adequacy (see Section 7).

Required packages: `nlme`; `ape`

To load our example data, download it from github.com/daveyfwright/ PCMsForPaleontologists and then follow the instructions in the file "Data_loading_script.R." To follow along, use the script in the file "full_script_PaleoPCM.R," or to see output without following along, see the file "example analyses.pdf." We include some of the output in the first two example analyses.

Is calyx shape correlated with fan density? To answer this, we could calculate the Pearson correlation coefficient, or perform an OLS regression.

Always plot the data, in this case `Shape` and `Density`, from the dataframe `allTraits`:

```
plot(allTraits$Shape, allTraits$Density,
    pch = 19, main = "",
    xlab = "Calyx shape ln(L/W)",
    ylab = "Fan density")
```

The standard test of correlation between two continuous variables is the Pearson correlation coefficient:

```
cor.test(allTraits$Shape, allTraits$Density)
```

Run this line of code and you should get the following output, which indicates shape and fan density are negatively correlated with p=0.02.

```
data: allTraits$Shape and allTraits$Density
t = -2.4878, df = 63, p-value = 0.01551
alternative hypothesis: true correlation is not
equal to 0
95 percent confidence interval:
 -0.50606523 -0.05952433
sample estimates:
      cor
 -0.2990814
```

The standard regression is OLS:

```
ols <- gls(Density ~ Shape,
           data = allTraits, method = "ML")
```

Look at the output:

```
summary(ols)
  Generalized least squares fit by maximum likelihood
    Model: Density ~ Shape
    Data: allTraits
         AIC      BIC     logLik
    175.9948 182.518 -84.99741


  Coefficients:
                  Value Std.Error   t-value p-value
  (Intercept)  3.561713 0.2087974 17.058224  0.0000
  Shape       -0.683057 0.2745674 -2.487756  0.0155
```

An OLS regression gives a slope of −0.68.

Perform PGLS regression to see if there is a difference:

```
tip.heights <- diag(vcv(phy=prunedTree))
cor.BM <- corBrownian(phy=prunedTree)
pgls <- gls(Density ~ Shape,
            correlation = cor.BM,
            weights = varFixed(~tip.heights),
            data = allTraits,
            method = "ML")
```

Note that this model uses two additional arguments that were not needed for OLS: (1) `correlation`, which defines the expected structure in the residuals, in this case based on the phylogeny and a Brownian motion model of evolution (explained in the next section), calculated using the function `corBrownian`; (2) `weights`, which is a modification to account for the different root-to-tip distances in our nonultrametric tree – important for paleontologists. Phylogenies that have extinct taxa in them and are scaled to time are referred to as "nonultrametric" because the tips end at different times. This is in contrast to "ultrametric" trees where all the tips end simultaneously at the present day, as is the case with trees of extant taxa.

Now that we have fit both OLS and PGLS, we can compare them using the AIC to see which is a better model for the data. AIC values can be used to evaluate the statistical fit of two or more candidate models, where the "best-fit" model is defined by having the minimum AIC score. To minimize overfitting, AIC provides a compromise between a model's goodness of fit to the data (i.e. the likelihood) and its "complexity" (i.e. number of parameters). For example, when there is low or no phylogenetic signal in the residuals of the regression, OLS would likely be preferred because it has fewer parameters. We can check the AIC scores of our data using the output of the gls function that we stored earlier in the objects `ols` and `pgls`.

```
summary(ols)$AIC
[1] 175.9948
```

```
summary(pgls)$AIC
[1] 212.8831
```

OLS has a lower AIC score, and we can therefore interpret our results as suggesting that OLS is a better model for the data. This in turn indicates that the phylogeny (assuming a Brownian motion model of evolution; see next section) does not explain much of the structure of the residuals of the regression. For further explanation, see the end of Section 3. It is useful to report both the OLS and PGLS regressions and AIC scores in a publication to allow readers to see the relative model fit.

It is possible to execute PGLS assuming a different model of evolution on the phylogeny, rather than Brownian motion. Such alternative models are discussed in the next section.

4 Tempo and Mode: Brownian Motion and More

Explicitly modeling morphological character change provides a framework with which we can quantitatively test hypotheses about macroevolution. Through models, we can link evolutionary histories of groups to different biotic or abiotic factors, and link patterns in morphological evolution to concepts like convergence, constraint, evolutionary radiations, and the adaptive landscape. In this section we will explore the basic models that form the foundation of many PCMs. The underlying principle here is to use mathematical models that describe how we expect a trait to change along lineages in a phylogeny, if evolution is operating in a particular way. The process of "model fitting" is essentially comparing the real data that we have for a trait at the tips of the phylogeny and adjusting the parameter values within the model (like the rate of evolution) to get the closest match possible between our real data and the expected possible trait values. Usually we are testing to find which model out of a set of candidate models is the one that is the best fit for our data (although see model adequacy in Section 7).

4.1 Brownian Motion

Brownian motion (sometimes known as a "random walk" model of evolution) is a model that can be used to describe the expected variance of a trait along branches of a phylogeny through time. It is called a random walk because the direction and amount of change in each time step is independent of the direction and amount of change in the previous time step. The average amount of change expected in each time step (the step rate, referred to by the model parameter sigma σ) is constant over time, so the trait variance across the tips of the phylogeny increases linearly over time. The simplest Brownian motion model assumes no inherent directionality of trait change (i.e. in each time step the trait value is equally likely to increase or decrease, so the expected net change is zero). In addition to the step rate σ, another component of the model is the phylogenetic trait mean (referred to by the model parameter theta θ), which is the estimated trait value for the ancestral node of the phylogeny.

Trait values of lineages wander around the phylogenetic mean through time. Under Brownian motion, the longer a branch's duration on the tree, the greater the *possible* divergence in trait change from the original value. However, since the direction of change under Brownian motion is equally likely to increase or decrease the trait value, the most probable value at any time is still the starting value. The range of possible trait values that a lineage could have can be visualized as an expanding cone through time; for Brownian motion the cone expands linearly and the center of the cone isn't expected to move through time (see figure 3.1 in Harmon 2019). When thinking about a whole clade, this pattern of change along lineages leads to a strong expectation – normally distributed trait values at the tips, with the standard deviation of that distribution dependent on the rate of evolution (the Brownian motion step rate σ). All else being equal, the morphological disparity of a clade will increase as a cone through time under Brownian motion (see figure 1 in Wesley-Hunt 2005; Erwin 2007). The less time that has elapsed since two branches split from a common ancestral branch, the more similar they are expected to be, because there has been less time for them to accumulate individual change and difference between them. We emphasize that the name "random walk," which is commonly used for this model, could cause confusion. On macroevolutionary timescales Brownian motion is unlikely to be random in the sense that it would result entirely from genetic drift (the true "random" microevolutionary process). Instead, the change over time is most likely as a result of the combination of microevolutionary forces, including selection, acting on the trait along with other mechanisms (e.g. drift, mutation, migration, etc.; see Hansen and Martins 1996). Importantly, the strength or direction of selection on a lineage might fluctuate over time as a result of different causes, for example adaptation to a stochastically varying environment, resulting in trait values sometimes increasing and sometimes decreasing, and therefore producing a statistically random pattern. Many underlying microevolutionary mechanisms are consistent with the macroevolutionary model for Brownian motion.

Despite the many-to-one mapping between microevolutionary processes and macroevolutionary outcomes, Brownian motion is almost always used in macroevolutionary model fitting because it can be thought of as a neutral model for trait evolution over long timescales. Moreover, because the equations underlying more complex models of trait evolution (e.g. Early Burst, Ornstein-Uhlenbeck; see subsection 4.2) often reduce to Brownian motion when their additional parameters approach zero, it is used as a kind of null model. It is often a useful starting point to compare the actual pattern of trait evolution to the simple Brownian motion expectation. If the best-fit model is Brownian motion, then no additional events or mechanisms (like a change in the rate of evolution or long-term constraints on trait change) are required to explain the pattern of macroevolution over time across the whole phylogeny. For other helpful explanations of Brownian motion on phylogenies see Nunn (2011) or Harmon (2019). See also the figures and equations in Butler and King (2004) for a clear, in-depth explanation of this and other models.

4.2 Ancestral State Reconstruction

A commonly used analysis in paleobiological studies is ancestral state reconstruction, for both continuous and discrete traits. This approach is a way of averaging trait values at tips, to make an estimate of what the trait value at a particular node might have been. Generally these estimates are very imprecise, and this large uncertainty around estimates means that as a standalone approach, especially for continuous traits, it is unlikely to provide many useful insights. To visualize this imprecision from one candidate ancestral value reconstruction approach, see the function `plotTraitgram` in `paleotree`. Furthermore, these approaches rely strongly on using a model of evolution to make the estimates; usually Brownian motion is assumed. It is possible to use other models (for example, Ornstein-Uhlenbeck; see next subsection) by weighting the average values according to the expectations of the model, but rarely are the adequacy of those models for the data assessed in advance. If the model assumed is not in fact a very good model for the data, then the ancestral state estimates will also not be accurate. For discrete traits the situation is a little better because possible states are constrained (see Section 8 for more detail). It is especially problematic to perform ancestral state reconstruction using a phylogeny of living taxa, without fossils; directional trait change through time or heterogeneity in tempo and mode across lineages will be missed. This will lead to inaccurate estimates of ancestral values (Finarelli and Flynn 2006).

In other parts of this Element, we refer to estimating "sampled ancestors"; this is unrelated to ancestral state reconstruction. Sampled ancestors are actual fossil specimens from which we can take measurements, and through probabilistic methods it is possible to estimate which of these is likely to be from a lineage whose descendants are also represented by at least one other actual fossil specimen in the dataset (Bapst 2013b, 2013a; Bapst and Hopkins 2017; Gavryushkina et al. 2017; Wright and Toom 2017). This is important because it is likely that many paleontological datasets contain at least one ancestor–descendant pair (Foote 1996), and ignoring this can distort the phylogenetic topology, branch lengths, and outcome of downstream analyses (Bapst 2014a; Soul and Friedman 2017). Sampled ancestors provide an actual trait value at a particular point in time along a lineage. In contrast, reconstructed ancestral states provide an imprecise estimate of a trait value for a hypothetical unsampled population from which two sister lineages descended.

EXAMPLE ANALYSIS 2 – FITTING A BROWNIAN MOTION MODEL

Required packages: `phytools`; `geiger`

A useful starting point in any phylogenetic investigation of morphology is to visualize the tree and trait plotted as a traitgram in "phylomorphospace." This means that the y-axis (see Figure 4) is meaningful, rather than when we usually plot trees where the branching order and position of the tips through time are important, but the position of the tips across the width of the tree is arbitrary. This can be done using the function `phenogram` from the package `phytools`. Traitgram and phenogram are words that are used interchangeably in the literature (Ackerly 2009). Positions of the tips on the y-axis are the measured trait values for taxa; the positions of the nodes are estimated assuming Brownian motion and therefore should not be considered rigorous estimates of ancestral trait values. At this stage they are just for visualization (see subsection 4.2). The function requires that the tree and trait have matching taxa and the trait vector has names corresponding to the tree tip labels (i.e. taxon names).

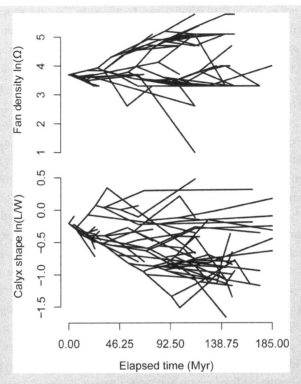

Figure 4 Phenograms, or traitgrams, help visualize patterns of trait evolution by projecting a phylogenetic tree into "trait space" (y-axis) and time (x-axis). The upper and lower diagrams were constructed using a model of Brownian motion to project the eucladid maximum clade credibility tree into the trait space of filtration fan density (upper) and calyx shape (lower), respectively.

Traitgram of Density
phenogram(prunedTree, Density, ftype = "off")
The argument ftype controls how the taxon names are displayed.

Traitgram of Shape
phenogram(tree, Shape, ftype = "off")

Now we will fit Brownian motion models to both of the continuous traits using the function fitContinuous from the geiger package.

```
densityBM <- fitContinuous(phy = prunedTree,
                           dat = Density,
                           model = "BM",
                           meserr = 0.12)
```

```
shapeBM <- fitContinuous (phy = tree,
                          dat = Shape,
                          model = "BM",
                          meserr = 0.03)
```

We can then look at the outputs densityBM and shapeBM, which show all the estimated parameter values like the step rate (the output "sigsq") and the phylogenetic mean (the output "z0").

densityBM

```
GEIGER-fitted comparative model of continuous data
  fitted 'BM' model parameters:
      sigsq = 0.012168
      z0 = 3.701690

model summary:
      log-likelihood = -66.448830
      AIC = 136.897659
      AICc = 137.091208
      free parameters = 2
```

shapeBM

```
GEIGER-fitted comparative model of continuous data
  fitted 'BM' model parameters:
      sigsq = 0.003723
      z0 = -0.206883

model summary:
      log-likelihood = -33.232143
      AIC = 70.464285
      AICc = 70.616184
      free parameters = 2
```

4.3 Beyond Brownian Motion

It is reasonable to think that a simple Brownian motion model of evolution might not be an adequate description of patterns of morphological variation over millions of years. As paleontologists we can think of many examples when patterns in the fossil record suggest periods of constraint, or rapid increases in morphological disparity. Fortunately, there is a straightforward framework that extends the Brownian motion model and can be used to model these kinds of patterns. The most commonly implemented alternative model is one that can be used to describe stabilizing selection – the Ornstein-Uhlenbeck model (Hansen 1997; Butler and King 2004; Hunt and Carrano 2010; Hunt 2012). This model includes three parameters. The first two are the step rate and the trait optimum identical to a Brownian motion model, but there is an additional parameter – alpha α. There is still a random walk component of trait change (included within the step rate), but each change is biased toward getting closer to an optimum trait value, and the strength of that bias or attraction toward the trait optimum is α. This is sometimes referred to as the elastic band parameter, because the further a trait value is from the optimum, the higher the strength of attraction to the optimum. The parameter α can be converted to phylogenetic half-life, which might be more straightforward to interpret. This is the time taken for a trait to evolve halfway to the new optimum and is calculated using $\ln(2)/\alpha$. The original application of Ornstein-Uhlenbeck in evolutionary biology was to paleontological data, highlighting the historic importance of paleobiology in establishing quantitative approaches to understanding macroevolution. It was applied to try to understand whether there had been an adaptive optimum in the evolution of tooth crown height in extinct equids (Lande 1976).

When modeling evolutionary change in a single trait along a lineage, the trait optimum might really represent an optimum value of high fitness for the organism (Hunt and Carrano 2010; Hunt 2012). Although this model was originally developed to describe stabilizing selection, when we are fitting an Ornstein-Uhlenbeck model to the evolution of a whole clade, the trait "optimum" likely represents some long-term average of trait optima that have varied through time. A useful metaphor for this could be a peak on a macroevolutionary landscape (see e.g. Boucher et al. 2018). The simple version of Ornstein-Uhlenbeck described here is sometimes referred to as a "single-peak" model, but further methods for modeling more complex macroevolutionary landscapes and/or shifts between peaks are available (discussed in subsection 9.1). It is also possible to include an extra parameter, which is the starting value for the trait at the root of the tree, which may be different to the trait optimum.

Brownian motion is a special case of the Ornstein-Uhlenbeck model where $\alpha=0$. Unlike for Brownian motion, for Ornstein-Uhlenbeck if $\alpha\neq0$, the variance does not increase linearly with time. After a particular time (related to α), the variance remains the same (see also figure 1 in Slater 2013). Traits evolving under Ornstein-Uhlenbeck are therefore less tightly linked to the structure of the phylogeny than traits evolving under Brownian motion. In other words, they will have less phylogenetic signal, and on average each tip value observed will be less similar to its nearest relatives than it would be expected to be if evolving under Brownian motion. An additional observation about Ornstein-Uhlenbeck is that if you have two sister tips, A and B, if the durations of the terminal branches leading to them are different (as for non-ultrametric trees), then the covariance of A with B will not be the same as the covariance of B with A, because covariance declines (i.e. phylogenetic signal is lost) with time since divergence.

If thought about further, it is possible to see how distinguishing one generating process from another might be difficult; a trait evolving under Brownian motion that is highly labile and therefore does not have a strong phylogenetic signal could show a similar trait distribution to the expectation under Ornstein-Uhlenbeck, as could a trait evolving under Brownian motion fit to a phylogeny that has errors in the relationships between taxa (for further explanation see Section 5). It is therefore important when fitting Ornstein-Uhlenbeck and other more complex models and interpreting the results to consider whether the quality of the input data is adequate to distinguish between these different scenarios. Despite these caveats, when extinct taxa are included in phylogenies that have been extensively researched to produce well-established phylogenetic relationships, the differences between expectations under different models are large enough that it should be possible to interpret model-fitting results with some confidence (Soul and Friedman 2015). An important advantage of including extinct taxa in a phylogeny is that it constrains possible trait change through time in ways that make it possible to distinguish between different models with confidence (Ho and Ané 2014b).

Aside from Ornstein-Uhlenbeck, examples of other models you may have heard of are "Early Burst" and "ACDC" (Blomberg et al. 2003; Harmon et al. 2010). These are models where the step rate σ changes exponentially through time. They are usually fit to data to try and understand the nature of adaptive radiations, for which theory suggests that you should see a high rate of morphological evolution early in the clade's history, or following a dispersal or extinction event, as morphological disparity accumulates through filling of new niches. The high early rate is followed by a period of stability characterized by a lower rate of morphological change (Slater 2013).

Another example of an alternative model is Brownian motion with a directional trend. This model is otherwise identical to Brownian motion except that each step change is consistently biased in one direction. This doesn't mean the change is only in that direction, but that there is a higher probability of a change one way than the other. It is not possible to detect evolutionary patterns consistent with directional Brownian motion with a tree of only living taxa; this model can only be applied to a tree that contains fossil data (Slater et al. 2012).

5 Incorporating Estimates of Error

5.1 Error in the Phylogeny

If there are errors in the tree topology (relative to actual evolutionary relationships), this will bias the analysis toward support for an Ornstein-Uhlenbeck model (Cooper et al. 2016), and in fact will generally bias support toward models with a low phylogenetic signal. This is because topological error reduces the apparent phylogenetic signal, making it appear as if trait change is less tightly linked to the structure of the phylogeny than the Brownian motion expectation. This effect cannot be corrected for within the analysis and is therefore something to be aware of when interpreting model-fitting outputs, especially for trees with fossil taxa.

5.2 Error in the Trait Data

The standard error of the sample and error in the trait measurements can also introduce bias. Noise from error can increase the apparent trait variance, leading to incorrect parameter estimates in model-fitting analyses, and a bias toward models with a lower phylogenetic signal, as is the case with error in the phylogenetic relationships. Often in paleontology we are only collecting measurements from a small number of specimens representing each taxon, which means that the standard error could be large. Silvestro et al. (2015) show with simulation that incorporating an estimate of error should be a priority, and that not doing so can bias toward Ornstein-Uhlenbeck as the best-fit model even when trait change was simulated under Brownian motion. Including an estimate of the standard error in any model-fitting analysis is therefore important (for a more thorough discussion of this in a paleontological context see Hunt and Carrano 2010). There are implementations of PCMs in R that allow inclusion of error estimates that make use of a variety of methods (e.g. Revell 2012). In Example Analysis 3 we have incorporated an estimate of standard error using the argument `SE` in the `fitContinuous` function.

6 Variation Across Trees

In the previous example analyses, each method is only applied to one tree topology. However, as mentioned in Section 1, it is *extremely improbable*

Required packages: ape; `phytools`; `geiger`

Simulations are a very useful tool to understand the behavior of different models and analyses (Barido-Sottani, Saupe et al. 2020). We can use simulation to get a better sense of what trait evolution on a phylogeny might look like under different macroevolutionary models.

The function `rtree` in `ape` is a very basic tree generation tool that gives nonultrametric trees with a particular number of tips, but doesn't use a divergence rate. For more sophisticated true simulation tools that can be used to generate trees under particular diversification or preservation scenarios, look at the packages `FossilSim` and `paleotree`. Other modeling packages (like `OUwie`) usually include tools that can be used to simulate characters on a tree under the particular models that the package implements.

Generate a tree with fifty tips
```
simtree <- rtree(50)
```

Simulate a trait evolving under a Brownian motion model of evolution on that tree and plot it using the function `bmPlot` from `phytools`

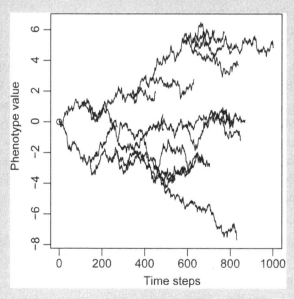

Figure 5 A single simulation of the changing values of one trait along the lineages of a phylogeny.

```
bmsim <- bmPlot(tree = simtree,
        type="BM",
        anc=0,
        sig2=0.01)
```

This function allows you to visualize a single character and how it changes through time along branches of the tree, in discrete time steps, as the clade diversifies (see Figure 5). Try changing the step rate using the argument `sig2` to see how that affects the trait distribution through time.

The output of `bmPlot` contains all the information needed to make the visualization; the simulated trait values at the tips, the tree, and all the trait values at each time step. We can also use this information to plot the trait values at the tips in a histogram. This matches the data we would have in reality and would be modeling. We would not be able to see the full trait history as in the visualization we just made, as we rarely know how a trait value has changed through time in all the lineages that make up a clade (although there are many studies of trait change through time in single lineages).

Plot a histogram of the trait values at the tips of the tree
```
hist(bmsim$x[1:50],
        xlab = "Trait values",
     main = "Simulated values at tips")
```

Other packages can also be used to simulate continuous trait change. The functions below simulate tip values for Brownian motion and Ornstein-Uhlenbeck using the `ape` package. Try changing the model parameters `sigma` and `alpha` (the step rate and constraint parameters). Try making the optimum value `theta` different from the starting trait value at the root. Use `phenogram` to visualize what effect these changes have, and to become more familiar with the kinds of patterns that will be best fit by the various different models.

```
BM <- rTraitCont(phy = simtree,
                    model = "BM",
                    sigma = 0.1)
phenogram(tree = simtree,
            x=BM,
            ftype = "off")
```

```
OU <- rTraitCont(phy = simtree,
                 model = "OU",
                 sigma = 0.1,
                 alpha = 0.1,
                 theta = 0,
                 root.value = 0)
phenogram(tree = simtree,
          x=OU,
          ftype = "off")
```

Below are scripts that can be used to fit models to real data and understand the output. The argument SE is used to input an estimate of the standard error (see Section 5 for further explanation).

First we will fit Ornstein-Uhlenbeck models to each trait using a function from the package geiger:

```
densityOU <- fitContinuous(phy = prunedTree, dat =
Density, model = " OU", SE = 0.12)
shapeOU <- fitContinuous(phy = tree, dat = Shape,
model = "OU", SE = 0.03)
```

Then we will fit an Early Burst model:

```
densityEB <- fitContinuous(phy = prunedTree, dat =
Density, model = "EB", SE = 0.12)
shapeEB <- fitContinuous(phy = tree, dat = Shape,
model = "EB", SE = 0.03)
```

Look at the parameter values for each model fit. Are they biologically reasonable? Checking parameter estimates is an important step, particularly because for Ornstein-Uhlenbeck models the different parameter estimates can interact with each other (Hunt 2012). Check that z0, the starting trait value, is within the range of tip values, and when investigating small changes on million-year timescales both alpha and sigsq should be small, on the order of 0.01 or less.

When you fit an Ornstein-Uhlenbeck model using geiger it should output a warning about using the VCV method. For our purposes, this warning can be safely ignored but deserves further comment. In the original version of this function, to fit Ornstein-Uhlenbeck quickly, the algorithm rescaled the branch lengths rather than operating on the VCV matrix directly (Slater 2013, 2014). This was problematic for nonultrametric trees because this particular kind of rescaling changed the covariance between tips in a way that assumed the input tree was ultrametric. The current version of this and other functions modify the VCV matrix directly to avoid the issue. Almost all PCMs (especially commonly used ones) can be applied to nonultrametric trees using R implementations, but this is a good reminder that it is important to check that none of the mechanics of the implementation are problematic when applied to paleontological trees, particularly with newly released method implementations.

Once the parameter estimates for the various models have been checked to make sure they are reasonable, the next step is to look at the AICcs (Akaike information criterion for small sample sizes) to find out which model is the best fit for the data and tree. The AICc is part of the output of the `fitContinuous` function that we used earlier, so it can be accessed using `densityBM$opt$aicc`.

These results indicate that Brownian motion is the best-fit model for `Density` but has a marginal AICc difference. Ornstein-Uhlenbeck is the best-fit model for `Shape`, with a larger AICc difference. Although most of the parameter estimates are biologically reasonable and broadly similar across different models, the exponential decline parameter in both Early Burst model fits is extremely small. In fact, it is at the lower boundary limits set by `geiger`. The maximum-likelihood estimate for this parameter is effectively zero for both traits and therefore the Early Burst model reduces to simple Brownian motion. However, note that the cost of this additional, unnecessary parameter leads to a decrease in model support.

that a single phylogeny (e.g. the maximum a posteriori tree from a Bayesian analysis, or a single most parsimonious tree) perfectly represents evolutionary "truth." Thus, it is important to get some measure of the robustness of results to variation in topology and branch lengths. We strongly recommend accounting for variation across phylogenetic hypothesis regardless of the methodological approach used to infer the tree topology and branch lengths; using a point estimate of the tree is never recommended. A simple way to understand how sensitive a particular result is to variation across trees is to run the analysis over a set of different trees. In Example Analysis 4 we randomly select 100 trees from the posterior distribution output by a joint estimation analysis. Other commonly used sizes for tree sets are 500 and 1,000, but really it is arbitrary and at the researcher's discretion as to how large a set of trees they need to include in order to fully characterize the variation in results.

EXAMPLE ANALYSIS 4 – APPLICATION TO MORE THAN ONE TREE

Required packages: `ape`; `geiger`

Here, we load the post-burnin posterior distribution of 1,000 trees and select a subsample of 100 of them, then make sure they all have a root age and no zero-length branches as described in subsection 2.1. Multiple trees are stored together in an object of the class "`multiphylo`," which can be treated as a list of trees. There are faster ways to run analyses over many trees (for example, using `apply` functions, or functions from the package `dplyr`, both of which we recommend), but here we use `for` loops so you can see clearly what is being done.

Load the post-burnin posterior distribution of trees:

```
trees <- read.tree("Eucladid_trees.tre")
```

Select 100 trees:

```
treeset <- trees[sample(1:1000, 100, replace = FALSE)]
```

The approximate age of the youngest taxon in the dataset is 268.8 Ma:

```
last.time <- 268.8
```

Adjust the root time; because all taxa are extinct, the `for` loop means that the root time for every individual tree in the tree set gets adjusted in the same way:

```
for (i in 1:length(treeset)){
    treeset[[i]]$root.time <- max(diag(vcv(treeset
    [[i]]))) + last.time
    }
```

Any of the previous analyses can be run over the entire tree set. For example, here we fit a Brownian motion model to all 100 trees using a for loop, to characterize variation in results.

Set up a vector to store results:

```
sigmas <- vector(length=100)
```

Run the model-fitting analysis on all 100 trees and store the evolutionary rate for each tree in the vector:

```
for(i in 1:100) {
    currentTree <- treeset[[i]]
    fitBM <- fitContinuous(phy = currentTree,
                          dat = Shape,
                          model = "BM",
                          SE = 0.03)
```

```
sigmas[i] <- fitBM$opt$sigsq
}
```
In this example we did not save the full output from the model fits, but rather just the parameter estimate for the step rate in each model fit. Decide beforehand what part of the results you need to access, and store them in vectors if they are single integers or lists if you want to store a more complex output.

Finally, view a histogram of all the estimates for the evolutionary step rate across the different tree topologies that we saved to the vector:

```
hist(sigmas)
```

If the phylogeny was not generated with a Bayesian inference approach that produces a posterior distribution of trees, other methods are available to account for uncertainties in tree topology or node ages. For example, if phylogenetic relationships were inferred using parsimony methods, one could select a random subset of the most parsimonious trees generated by the analysis (Lloyd et al. 2012). Similarly, you could choose to investigate the sensitivity of your results across a set of timescaled trees from the output of a stochastic timescaling algorithm (e.g. Bapst 2013a) to understand the effect of branch-length variation on your analysis.

Once analyses have been run over an entire distribution of trees, the results should be summarized appropriately. Unfortunately, no universally agreed-upon way to do this has been developed. Nevertheless, many reasonable approaches are possible for visualizing and summarizing the results. For example, displaying median values with the variation around them (as long as it is made clear that this variation is separate from the error or confidence intervals of the results themselves), showing model support across trees in a bar plot (figure 6 or figure 2 in Halliday and Goswami 2016), or showing a representative result and discussing how robust it is have all been used in the literature and are improvements over reporting a point estimate alone (for more examples see Hopkins and Smith 2015; Clarke et al. 2016; Soul and Friedman 2017). Where appropriate for the analysis, we suggest reporting the distribution of p-values across a sample of trees in the form of its frequency distribution and summary statistics. For example, some authors have summarized results of phylogenetic statistical tests in terms of either the mean p-value estimated across the distribution of trees and/or the proportion of samples where the test statistic exceeds some a priori threshold of statistical significance (Soul and Benson 2017; Cole et al. 2019). A subset of PCM implementations do have built-in ways of summarizing across sets of trees (see for example rjmcmc in geiger), but this is rare. The most important thing is to choose an approach that makes the variation in results clear, as reporting only the average is misleading unless there is truly no variation across trees. Note that if there is a large amount of variation in relationships across topologies (i.e. the phylogeny is poorly constrained), then in model-fitting exercises there is likely to be a bias toward support for models with a low phylogenetic signal, for the reasons discussed in section 4.3. The combination of variation across topologies and support for a low-signal model should lead to skepticism about whether there is enough information in the data to reliably distinguish between models.

7 Model Fit and Model Adequacy

The previous Example Analyses 1–4 involved fitting models of evolution to data by estimating the maximum-likelihood parameter values for a particular model given the data, then comparing the AIC scores of each model fit to see which was the best-fit model, and what that might tell us about evolution of the clade in question. However, just because one model is a better fit than the other models that were tested, it does not necessarily mean that it is actually a good model for the data (Pennell et al. 2015; Voje et al. 2018). No model can encompass all the different processes that lead to the actual pattern of trait change, but the important thing to consider is whether a particular model captures the evolutionary dynamics to the extent that it allows you to answer the question you are interested in addressing. In this section we explore how to approach testing model adequacy. There is not a single test that can say definitively, "this model is good enough," but there are ways to understand how much information is in a dataset, how reliable the outcome of a model-fitting exercise is, or how confident we should be in the answer to an evolutionary question derived from PCMs. Often, finding the ways in which a model does not capture aspects of the data (i.e. model misspecification) can provide extra information about evolutionary processes.

There are a variety of reasons that simple models of trait evolution applied across a whole clade might not be good models for the underlying dynamics (see e.g. Polly 2019). Furthermore, it may be the case that there is simply not enough information in your dataset to allow you to identify an adequate model, especially if the model you are interested in assessing produces similar patterns of trait evolution to other candidate models (Boettiger et al. 2012; Slater and Pennell 2014). With simple models like linear regression, taking steps such as plotting the data, applying goodness-of-fit tests, or inspecting the residuals can help establish whether the model is adequate. For more complex approaches (i.e. most PCMs) visualization alone does not often help (Revell et al. 2018). There are no well-established model adequacy testing approaches for PCMs on nonultrametric (i.e. fossil) trees. Even for analyses on trees of living taxa, model adequacy testing is still relatively uncommon, despite growing recognition of its importance within the PCM literature.

Throughout this Element we mention the possibility of the dataset not containing enough information to make reliable inferences based on model-fitting outcomes. Simulation is one approach that can be used to test whether or not this is the case. For example, posterior predictive simulation can be used to compare the relative adequacy of two candidate models and allow greater power to be able to distinguish between them (see Slater and Pennell 2014). A more general simulation framework can be used to investigate the power to reliably identify a particular model. We suggest the following steps as a straightforward addition to model-fit comparisons, which are made possible by the wide range of simulation tools now available in R:

(1) Find the best-fit model (or models if they have similar goodness of fit).
(2) Identify the fitted model parameters.
(3) Use these parameters to repeatedly simulate trait change under the best-fit model(s) on the original phylogeny.
(4) Perform model fitting and parameter estimation on each of the simulated datasets.
(5) Record how frequently the model that was used for simulation is correctly identified.

If the model that was used for simulation is not often identified as the best-fit model, or the fitted parameter values vary widely across simulation runs, then this suggests there is not enough information in the dataset to reliably inform inference, and the original outcome should be interpreted cautiously.

For a subset of phylogenetic comparative approaches it is possible to use standardized phylogenetic contrasts to investigate model adequacy and identify types of model misspecification (Pennell et al. 2015). One statistical condition of the phylogenetic contrasts of observations of a continuous trait on a phylogeny, under some models, is that they are independent and identically distributed (IID). This is true for nonultrametric trees when a Brownian motion model of evolution is assumed, as well as for many related Gaussian models, even where the mode or rate of evolution changes across the tree, but it is not true for any kind of Ornstein-Uhlenbeck model on a nonultrametric tree (although it is possible to perform a scalar transformation of the branch lengths that results in IID contrasts; see Ho and Ané 2014a). If the contrasts are IID, then various test statistics on the contrasts can be used. The procedure (outlined in detail in Pennell et al. 2015; see table 1 of that paper in particular for details of the types of model misspecification that can be identified with different test statistics) is similar to that used earlier, except that the test statistic of interest is applied to the contrasts of the real data and the contrasts of each of the simulated datasets. If the test statistic value for real data falls outside of the distribution of the test statistic across the simulated datasets, then there is some degree of misspecification, or inadequacy, of the model. For example, if the mean of the squared contrasts is not within the simulated distribution, this indicates that the overall rate of evolution is over- or underestimated.

8 Post-Hoc Modeling of Discrete Characters

The majority of models we have discussed thus far are appropriate only for continuously valued traits. However, discrete morphological characters have been particularly important in paleontology because they comprise the bulk of primary data for inferring phylogenies of fossil taxa, including both parsimony and model-based approaches (see Wiley and Lieberman 2011; Wright 2019; Warnock and Wright 2020) and investigating patterns of morphological disparity (Lloyd 2016). In addition to intrinsically discrete traits (e.g. presence/ absence), many other kinds of morphological, paleoecological, and taphonomic traits are often categorized into discrete character states for practical reasons (e.g. species occurrences in reef vs. non-reef settings).

Table 1 Model-fitting results for analyses of trait evolution in eucladid crinoids described in Example Analysis 3

Trait	Model	logL	AICc	Parameter estimates
Filtration fan density	**BM**	**−66.449**	**137.09**	$\sigma^2 = 0.012$, $z_0 = 3.70$
	OU	−66.250	138.89	$\alpha = 0.002$, $\sigma^2 = 0.013$, $z_0 = 3.71$
	EB	−66.449	139.29	$a = -0.000001$, $\sigma^2 = 0.012$, $z_0 = 3.70$
Calyx shape	BM	−33.209	70.57	$\sigma^2 = 0.0037$, $z_0 = -0.207$
	OU	**−29.337**	**64.98**	$\alpha = 0.0094$, $\sigma^2 = 0.0050$, $z_0 = -0.333$
	EB	−33.209	72.73	$a = -0.000001$, $\sigma^2 = 0.0037$, $z_0 = -0.207$

Models with the best fit according to AICc scores (Akaike information criterion for small sample sizes) are indicated in bold. BM = Brownian motion; OU = Ornstein-Uhlenbeck; EB = Early Burst.

Required packages: `ape`; `OUwie`

In Example Analysis 3 we found that Ornstein-Uhlenbeck was the best-fit model out of those that we tested for the `Shape` data. Here we examine whether that Ornstein-Uhlenbeck model was actually a good model, and reliably identifiable from the data. To do this we will need to simulate under an Ornstein-Uhlenbeck model of evolution; the package `OUwie` has a function that does this.

First, extract the parameter estimates from the previous model-fitting output:

```
alpha <- shapeOU$opt$alpha
sigma.sq <- shapeOU$opt$sigsq
theta <- shapeOU$opt$z0
```

Prepare the data so that they are in the format that can be read by `OUwie` (see the `OUwie` help file for further details).

```
shapeData <- data.frame(species = names(Shape),
                        regime = 1,
                        shape = Shape)
```

Use the function `replicate` to repeatedly simulate under Ornstein-Uhlenbeck with the same parameters and store the simulated trait values for the tips of the tree. The function `replicate` is a convenient shortcut; it does the same thing as a `for` loop but is faster when you want to repeat the same operation many times. The simulated tip values are output in the variable X by the function, so this is the element we need to store.

```
simtrait <- replicate(100,
        OUwie.sim(phy = tree,
            data = shapeData[,1:2],
            root.age = tree$root.time,
            alpha = c(alpha,alpha),
            sigma.sq = c(sigma.sq,sigma.sq),
            theta = c(theta,theta),
            theta0 = theta
            )
        $X)
```

Set up the simulated trait values into vectors with names that can be read by the model-fitting function.

```
form.simtrait = list()
for(i in 1:100) {form.simtrait[[i]] <- simtrait[,i]
                names(form.simtrait[[i]]) <- names
                (Shape)
                }
```

Fit the single-peak Ornstein-Uhlenbeck model, which was the best-fit model for our data before out of the models we tested (warning: if you do this over 100 iterations, it will take a long time).

```
OUfits <- lapply(X = form.simtrait,
                FUN = fitContinuous,
                phy = tree,
                model = "OU")
```

Fit a single-rate Brownian motion model to the same sets of simulated trait data:

```
BMfits <- lapply(X = form.simtrait,
                FUN = fitContinuous,
                phy = tree,
                model = "BM")
```

If you run this analysis using the eucladid tree, you should get several warning messages during the Ornstein-Uhlenbeck model fitting to simulated data, a first indication that the resulting model fit is unlikely to be reliable. If you inspect the output model-fit AICcs and estimated parameters, you should find that Ornstein-Uhlenbeck is the preferred model over Brownian motion, but that the estimated parameters for Ornstein-Uhlenbeck are not similar to the parameters used for simulation, and are quite high. This suggests that the best-fit Ornstein-Uhlenbeck model is unlikely to be adequately capturing variation in the trait data. A next step might be to investigate whether a single-peak Ornstein-Uhlenbeck model is missing heterogeneity in rate or mode of evolution (see Section 9).

In practical terms, if the discrete trait of interest is included directly in a model-based phylogenetic inference analysis, it is possible to jointly estimate transition rates between states and ancestral states with other parameters in a Bayesian context (Warnock and Wright 2020). However, it is far more common to investigate patterns of discrete trait evolution across a phylogenetic tree (or set of trees) *after* conducting a phylogenetic inference analysis. Such post-hoc modeling typically involves the estimation of ancestral states (i.e. the trait values at the nodes of the tree), which can then be used to characterize the timing and phylogenetic position transitions between states. In a similar way to the approach we showed earlier for continuous traits, fitting alternative models of discrete trait evolution in R can be done with functions in the `geiger` package.

One of the methods most widely used to estimate ancestral states for discrete characters is parsimony. Given a tree depicting evolutionary relationships, parsimony-based reconstructions map a trait onto the tree such that the number of evolutionary changes is minimized (Maddison and Maddison 2020). Parsimony methods do not incorporate information regarding branch durations, so changes are equally likely to occur on short and long branches. Notably, parsimony-based reconstructions cannot always be unambiguously determined, so multiple parsimonious reconstructions are sometimes possible, which can lead to equivocal results. Beyond parsimony methods, modeling discrete characters is most commonly performed using a variation of the Mk model described by Lewis (2001), which involves a single parameter: the rate of evolutionary change. In contrast with parsimony, the Mk model involves estimating branch lengths, and allows for the possibility of multiple changes to occur along a branch, such that the total number of evolutionary transitions can be greater than the minimum required under parsimony. In its simplest form, the Mk model assumes changes are equally likely between character states, and these states have identical equilibrium frequencies. However, in a Bayesian context these assumptions can be relaxed to accommodate a wide variety of possible model variations, including rate variation among characters (Wagner and Marcot 2010; Wright 2019), unequal transition frequencies between state changes (Wright et al. 2016), and even whether ecological vs. nonecological trait partitions evolve at different intrinsic rates (Wright et al. 2020).

Although simulation-based studies show Bayesian methods using the Mk model often outperform parsimony in phylogenetic inference (Wright and Hillis 2014; O'Reilly et al. 2016; Puttick et al. 2019), both methods perform well when reconstructing ancestral states given a tree (Gascuel and Steel 2014; see next paragraph), and the decision to choose parsimony vs. model-based approaches may depend on your research question. For example, a researcher might be interested to show that a particular phylogenetic hypothesis requires some minimal threshold of evolutionary changes; in such cases, parsimony methods may be preferred because they always prefer the minimum number of changes. In that sense, parsimony is biased to underestimate the total number of character changes when the rate of evolution is high; however, the extent to which a researcher can use that bias to their advantage may be useful for some types of questions (i.e. "this character changed *at least* X times across the phylogeny"). Moreover, sometimes it may be of interest to compare the minimum number of changes required between different phylogenetic hypotheses (regardless of how the trees were inferred). Similarly, parsimony methods (and related approaches) for discrete character evolution can also be used to study rate variation and patterns of character evolution (Wagner 2012), and inform rate distributions in model-based phylogenetic methods (Harrison and Larsson 2015; Wright 2017a).

Unlike reconstructing ancestral states for continuous traits (see subsection 4.2), methods to reconstruct ancestral states for discrete characters are remarkably robust and provide far more accurate estimates than their continuous trait counterparts. For example, a simulation study by Gascuel and Steel (2014) shows both parsimony and model-based methods of discrete ancestral state reconstruction perform well, have nearly the same accuracy, and are robust to sampling bias and model misspecification. As discussed for continuous traits, the fidelity of methods to reconstruct ancestral state values for discrete characters improves when fossil data are incorporated into the tree. Puttick (2016) used simulations to examine the impact of including fossil taxa in analyses of discrete trait evolution in extant lineages, and showed that including both fossil and extant taxa consistently outperformed analyses with extant taxa alone. In practice, there could be errors in coding derived from incomplete or fragmentary fossil material, or perhaps incorrect assignment of a habitat based on incorrect paleoenvironmental reconstructions (e.g. freshwater vs. marine). Puttick (2016) therefore also investigated the impact of including a proportion of fossil taxa having *incorrect* discrete trait values and found that analyses of fossil and extant taxa still outperformed extant-only analyses at estimating ancestral states, even when up to 75 percent of fossil taxa were incorrectly coded. These results highlight the important role of fossil taxa for estimating ancestral states for discrete characters, and encourage the inclusion of as many fossil taxa as possible, even when there is uncertainty in their underlying traits.

Another popular model-based method for post-hoc modeling of discrete character evolution is a Bayesian approach called stochastic character mapping (Nielsen 2002; Revell 2013). In this method, discrete trait evolution is simulated an arbitrarily large number of times across a tree, and the resulting character histories are sampled proportional to their posterior probability, using Markov-chain Monte Carlo, to generate the posterior distribution. Although the output can be difficult to visualize for traits with more than two states (Revell 2013), a benefit of stochastic character mapping is that the

Required packages: ape; `phytools`; `geiger`

In this example we will fit three models of discrete character evolution to the MCCT for eucladid crinoids to investigate the evolution of calyx complexity, an important trait related to growth patterns and rates of development (Kammer 2008; Wright 2015; Ausich et al. 2020). Calyx complexity is defined as the number of plates interrupting the posterior interray (more plates = increased "complexity"). This trait typically varies from zero to three among most Paleozoic eucladid taxa, and is therefore a multistate character with four discrete states.

First, let's fit a model with a single-rate parameter describing all possible transition rates between character states. For example, if a character has two discrete states (e.g. number of posterior plates), labeled "1" and "2," this single-parameter model assumes the transition rate between states "1" ➜ "2" is equal to the backward transition between "2" ➜ "1." The same logic can be extended to multistate characters, such as our example of calyx complexity. This model is specified as equal rates, or "ER" for the model argument in the fitDiscrete function.

```
ER <- fitDiscrete(tree, dat = Complexity,
model = "ER")
```

In contrast with the simplistic equal-rates model, we can also specify a more complex model where each possible rate is a unique parameter. For example, a different rate is estimated for transitions between states "1" ➜ "2" and "2" ➜ "1," etc. In `geiger`, this model is specified as all rates different, or "ARD" for the model argument in the fitDiscrete function.

```
ARD <- fitDiscrete(tree, dat = Complexity,
model = "ARD")
```

For discrete traits with more than two states, we might predict that certain kinds of traits evolve along an order sequence of changes between states. For example, based on developmental patterns, some crinoid researchers consider that calyx complexity should be an ordered character; that is, transitions should be restricted so that they can only occur between consecutive states (e.g. Wright 2015, 2017). In this case you can use the meristic model to model transition rates.

```
MER <- fitDiscrete(tree, dat = Complexity,
model = "meristic")
```

As in previous examples, models can be compared using the AICcs.

```
ER[[4]]$aicc
ARD[[4]]$aicc
MER[[4]]$aicc
```

Here, we have found that the meristic model corresponding to ordered state transitions is the best-fit model for the data.

It is also straightforward to reconstruct ancestral states, and the associated uncertainty, under an equal-rates model. phytools has useful functions for plotting the likelihood of a given state onto the tree.

First, estimate ancestral states using the function ace from the ape package:

```
anc.char <- ace(Complexity, tree, type = "discrete")
```

Ladderize the tree for better visualization and then plot in the usual way:

```
plot(ladderize(tree), show.tip.label = FALSE)
```

Add pie charts of character state likelihoods to the nodes:

```
nodelabels(pie = anc.char$lik.anc,
           piecol = c("lightblue", "darkblue",
           "salmon", "darkred"),
           cex = 0.5
           )
```

uncertainty in patterns of character evolution can readily be accounted for by examining the posterior distribution of character histories. As might be expected, this method presents a challenge for any attempts to summarize results across a distribution of time-calibrated trees (see Section 6 for ideas regarding how to summarize results across tree distributions). Stochastic character mapping can be conducted in R using the make.simmap function in the package phytools.

A common feature of models discussed thus far, for both continuous and discrete traits, is that the underlying dynamics of trait evolution do not change over time and/or among clades. Relative to the literature on discrete traits, modeling heterogeneous dynamics of trait evolution is far more developed for continuous characters, which is more thoroughly discussed in Section 9. However, an important, promising family of methods for investigating heterogeneous dynamics in discrete character evolution was developed by Lloyd et al. (2012). Using discrete character matrices (such as those constructed for phylogeny estimation or disparity analyses), these approaches use likelihood ratio tests to evaluate whether a given phylogeny is consistent with single-rate vs. multi-rate models, where two (or more) rates may arise from either temporal shifts in evolutionary dynamics and/or along specific branches. This method attempts to account for patterns of character sampling and completeness, and has been applied to a variety of macroevolutionary questions in a variety of fossil groups, including lungfish (Lloyd et al. 2012), echinoids (Hopkins and Smith 2015), Mesozoic mammals (Close et al. 2015), and Paleozoic crinoids (Wright 2017b). Relevant functions for performing these kinds of analyses are available in the R package `Claddis` and further described by Lloyd (2016).

9 Modeling Heterogeneous Trait Evolution

In this section we provide descriptions and some introductory scripts for a variety of more complex, or recently published, approaches. These approaches deal with modeling continuous trait evolution where the tempo or mode varies through time or across lineages (for examples of approaches that can be applied to discrete trait evolution to model heterogeneous change, see Section 8). This is not an exhaustive review of all the approaches available, but a summary of some of those that might be of most interest to paleontologists, and references where they have been applied in a paleobiological context.

When starting out with more complex models it is appropriate to be cautious. There are several publications that highlight situations where phylogenetic comparative approaches do not behave as expected or intended (Freckleton 2009; Uyeda et al. 2018). It is very important to ensure that any method applied is appropriate for the question and data, and that it is applied correctly. This is likewise true of all the more straightforward analyses we explained earlier, but it can be more difficult to identify problems as models become more complex. Like all statistical methods, PCMs have assumptions, biases, and limitations (Cooper et al. 2016). Because PCMs comprise a diverse ensemble of models, we cannot provide a comprehensive guide or checklist of assumptions pertaining to particular methods here. Instead, we emphasize it is critical for researchers to carefully consider the statistical assumptions of a given PCM (and their biological implications) when designing a study or interpreting results. We reiterate that this advice is general to all biological sciences and not an issue specific to PCMs.

In fact, many criticisms of PCM approaches can be addressed by a researcher (1) ensuring they fully understand the method being implemented, as well as the data it is applied to, to identify possible biases or features of the data that could lead to misleading results (e.g. sampling rate; Soul and Friedman 2017); (2) taking important steps such as checking that output parameter estimates are biologically reasonable – if in doubt, simulation can be a practical way to try and characterize the behavior of a particular method when applied to data similar to your own (Barido-Sottani, Saupe et al. 2020); and (3) not overinterpreting results. Finally, we note that many potential issues with PCM-based studies are alleviated or otherwise improved when fossil taxa are included (Slater et al. 2012; Ho and Ané 2014b; Hunt and Slater 2016).

9.1 Multi-Regime Ornstein-Uhlenbeck Models

Ornstein-Uhlenbeck models have three or four parameters that define an "adaptive peak" on a phenotypic landscape (sensu Simpson 1944). In the context of phylogenetic paleobiology, this peak can also be thought of as a macroevolutionary regime, or macroevolutionary landscape (Boucher et al. 2018). For some period of time, the evolution of a set of taxa can be characterized using single estimates for each model parameter, and though the actual mechanisms generating the observed patterns may vary through time, they have the same outcome. This model is consistent with Simpson's concept of the adaptive landscape. Often, we would not expect that a clade would remain in the same macroevolutionary regime (i.e. where the same trait value is optimal and the strength of α [the constraint parameter] remains the same) over the many tens of millions of years of its evolutionary history, or that all branches of a clade would experience the same regime. We might also be interested in how the regime a clade experiences relates to factors like environmental change, or whether multiple nonsister lineages experienced the same regime, which can result in convergence.

It is possible to model shifts in the evolutionary regime through time, or along lineages, by estimating new parameter values for different sections of the phylogeny. The package OUwie has model-fitting and trait simulation functions for this purpose. OUwie has several options for models; you can allow the strength of the selection parameter α, the trait optimum θ, or the evolutionary rate σ to shift. You can also fix or vary different combinations of these for each regime. The function requires a priori definition of clade or time regime shift points; it is therefore suitable for hypothesis testing (Gearty and Payne 2020). When choosing shift points for an analysis, they should be related to specific hypotheses about factors that may have affected the evolutionary history of the clade (e.g. a novel adaptation, dispersal to a new region, or environmental changes). Thus, we emphasize it is *critical* to have in-depth knowledge of the taxonomic group under investigation to meaningfully address evolutionary questions with these methods. If you are not an expert on the group you want to investigate, we encourage you to consider reaching out to collaborate with one. A variety of other packages can also be used to investigate multi-regime models (including multivariate multi-regime models) of trait evolution, including mvMorph, mvSLOUCH, OUCH, geiger, and ape.

Multi-regime Ornstein-Uhlenbeck models can be very complex and are therefore often computationally difficult to fit. It is especially important to check for convergence when estimating parameters (Gearty and Payne 2020) and that the parameter values are sensible. For example, for OUwie, ensure that the results show estimated values for θ (theta_0, the value at the root of the tree) that are within or close to the known range of trait values, and that the step rate is not unreasonably high. If it is not clear what could be considered a reasonable rate, it should usually be possible to find rate estimates in the literature for the same taxon, or for similarly preserved taxa.

9.2 Character-Dependent Continuous Trait Change

A further question that can be answered with multi-regime Ornstein-Uhlenbeck models is how a particular discrete character has influenced the evolution of other continuous traits. OUwie can be used to test this. If the discrete trait is confined to particular clades, then the same approach as in Example Analysis 7 can be used to define regimes. If the discrete trait is distributed over the tree rather than being confined to particular clades, then you can use any ancestral state reconstruction approach to reconstruct the state at each node (see Section 8), and then use the different estimated states to define the regimes on the tree. Uncertainty in the ancestral states estimated can be appropriately accounted for using the package SIMMAP. The additional layer of uncertainty derived from estimating the ancestral states under a particular model of evolution should be taken into account when interpreting results. Price (2019) provides a worked example for paleontological data (see also Price et al. 2015).

9.3 Heterogeneous Rates on Individual Branches

It is possible to model change of one or more continuous traits on a phylogeny with rate differences between and/or along each branch, and a range of models has been developed to do so (O'Meara et al. 2006; Hunt 2013; Revell 2013; Landis and Schraiber 2017). Markov chain Monte Carlo (MCMC) approaches are computationally well suited for use to fit this kind of highly heterogeneous model. One example model and Bayesian implementation is AUTEUR (Eastman et al. 2011), which is implemented with the function rjmcmc.bm

EXAMPLE ANALYSIS 7 – MULTI-REGIME ORNSTEIN-UHLENBECK

Required packages: OUwie

Most of this example script serves to correctly set up the input objects. First, set up a dataframe and tree in the format that they can be read in. OUwie has a very specific format required for the input data. It requires a three-column dataframe where the first column contains the taxon names, the second contains the regime each tip belongs to (at the start the regime labels are placeholders for when we define them according to a hypothesis later), and the third column contains the trait data. You can optionally give a fourth column with the standard error of the trait measurements (this means an estimate of the intraspecific error if you have multiple specimens, plus measurement error; see Section 5 on error) – this is advisable whenever possible. The tree must have internal node labels corresponding to the regime the node falls within. At the time of publication it is the case that to prevent the function outputting an error message, there must be two or more regimes, even if a single-regime model is subsequently fit.

Here we use the same crinoid data for the Density trait that we have been using previously. The possible regime shift points defined further down are based on DFW's prior knowledge as a crinoid systematist and taxonomist familiar with their morphology and ontogeny, derived from previous experience and knowledge of the published literature.

Set up a dataframe for the trait we are using, with three columns as specified earlier:

```
densityData <- data.frame(species = names(Density),
                          regime = 1,
                          density = Density)
```

Add node labels to the phylogeny we have been using

```
prunedTree$node.label <- sample(c(1,2),
                                Nnode(prunedTree),
                                replace = TRUE)
```

Once the data and phylogeny are correctly formatted, we can estimate parameters for single-rate Brownian motion and Ornstein-Uhlenbeck models, similarly to how we demonstrated in Example Analysis 3 using the alternative R package geiger. The root age of the tree must be provided, and it can be helpful to provide starting values for the parameters to aid convergence. Ideally the algorithm should be started from a variety of values to make sure that it converges on the same parameter estimates, and not local optima.

By default, the algorithm does not estimate theta_0, the trait value, at the root as part of the model. Instead, the optimum at the root is dropped, which can stabilize estimates of the adaptive optimum trait value. If there is reason to include estimating theta_0, be aware that sometimes estimates of theta_0 and alpha can interact (Ho and Ané 2014b), so it is important to check for sensible parameter estimates after the model fit.

First, fit single-regime Brownian motion and Ornstein-Uhlenbeck to the Density trait to provide a null against which more complex models can be compared.

```
densityBM <- OUwie(phy = prunedTree, data = densityData,
                   model = "BM1",
                   quiet = TRUE,
                   root.age = tree$root.time)
densityOU <- OUwie(phy = prunedTree, data = densityData,
                   starting.vals = c(0.01, 0.01),
                   quiet = TRUE,
                   model = "OU1",
                   root.age = tree$root.time)
```

Now we can specify the node labels on the tree according to the hypothesized regimes. There are two subclades within the Eucladida tree that seem to show many modifications associated with building a dense filtration fan (e.g. many branches per arm or many pinnules on branches). OUwie can be used to test whether this high fan density was associated with a different trait mean or rate of evolution. In this example we will define the tip and node regimes ourselves beforehand because it involves more than one clade. If we hypothesize a rate shift in only one clade, it is easiest to use the "clade" argument within the OUwie function rather than defining the regimes beforehand.

Find the node defining the first clade of interest:

```
mrca1 <- getMRCA(prunedTree, c("Blothrocrinus",
"Hydreionocrinus"))
```

Make a vector of all nodes and tips that descend from it:

```
first <- c(mrca1, getDescendants(prunedTree, mrca1))
```

Do the same for the second clade:

```
mrca2 <- getMRCA(prunedTree, c("Pirasocrinus",
"Zeacrinites"))
```

```
second <- c(mrca2, getDescendants(prunedTree, mrca2))
```

Combine the two vectors to a single one, specifying all nodes and tips that we hypothesize belong to the "high-density fan" regime:

```
desc <- union(first, second)
```

Separate out node and tip vectors and label the nodes of the phylogeny with regimes accordingly:

```
tips <- prunedTree$tip.label[desc[desc<=Ntip
(prunedTree)]]
nodes <- desc[desc>Ntip(prunedTree)]-Ntip
(prunedTree)
treeD <- prunedTree
treeD$node.label <- rep(1, Nnode(treeD))
treeD$node.label[nodes] <- rep(2, length(nodes))
```

Modify the trait dataframe to specify the correct regime labels for tips:

```
regimetableD <- densityData
regimetableD[tips,2] <- 2
```

Now that the data and phylogeny are prepared, we can fit a variety of multi-regime models to the data. Any combination of model parameters can be fixed or allowed to change between regimes, using the `"model"` argument (refer to the documentation for OUwie for a full explanation of all the options for this argument and what they mean). Here we did not include estimating `theta_0`. Depending on the number of taxa in the dataset and the variation in trait values, very complex models such as one where all parameter values change (`"OUMVA"`) may not be well fit to the data or generate reliable parameter estimates. In that situation, variation across different possible topologies will be useful to assess how reliable the result is. Below is code to fit a single-regime Brownian motion model, and an Ornstein-Uhlenbeck model with two regimes that have different trait means. The other models can be fit by modifying the model argument. See the full script online for the code to do this. The results of fitting all of the available models are shown in Table 2.

```
high_densityBMS <- OUwie(phy = treeD, data =
                         regimetableD,
                         model = "BMS",
                         quiet = TRUE,
                         root.age = tree$root.time)
```

```
high_densityOUM <- OUwie(phy = treeD, data =
regimetableD,
                          starting.vals = c(0.01, 0.01),
                          model = "OUM"
                          quiet = TRUE,
                          root.age = tree$root.time)
```

After parameter estimates are checked, AIC$_c$s can be compared to those for the simple model. Although OUwie may indicate that a reliable solution to the model fit has been reached, it is common for more complex models like OUMVA to have unrealistic estimates of parameters, indicating the algorithm did not effectively fit the model. In fact, this is the case with the eucladid data we are using, where very high model support for OUMVA only occurs in the case where parameter estimates are biologically unrealistic (Table 2).

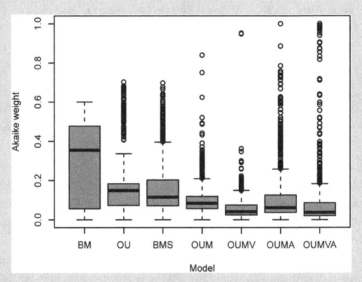

Figure 6 Mean and variation in Akaike weights for seven candidate models of continuous trait evolution. Note that although the mean weight for complex models is low, for some trees within the set, complex models have high weights. These often represent fits with unrealistic model parameters.

A single-rate Brownian motion model is the best-fit model but it is only marginally better than the next best model, or in fact than most of the other models (Table 2). It is likely the case that there is not enough information in our dataset to be able to meaningfully distinguish between these models. Figure 6 is an example of how to summarize variation across a set of trees; we show the median and variance of Akaike weights for each model when it has been fit to data combined with 100 trees drawn randomly from the posterior distribution.

Table 2 Model-fitting results for multi-regime Ornstein-Uhlenbeck analyses

Model	logL	AICc	Parameter estimates
BM	−66.449	137.09	$\sigma^2 = 0.012$, $\theta = 3.70$
OU	−66.250	138.89	$\alpha = 0.002$, $\sigma^2 = 0.013$, $\theta = 3.71$
BMS	−64.831	138.33	$\sigma^2_1 = 0.011$, $\sigma^2_2 = 0.012$, $\theta_1 = 3.68$, $\theta_2 = 5.48$
OUM	−64.442	137.55	$\alpha = 0.0023$, $\sigma^2 = 0.0023$, $\theta_1 = 3.68$, $\theta_2 = 5.51$
OUMV	−64.474	139.96	$\alpha = 0.0016$, $\sigma^2_1 = 0.0373$, $\sigma^2_2 = 0.0354$, $\theta_1 = 3.68$, $\theta_2 = 5.50$
OUMA	−64.421	139.86	$\alpha_1 = 0.0027$, $\alpha_2 = 0.0053$, $\sigma^2 = 0.0126$, $\theta_1 = 3.68$, $\theta_2 = 5.03$
OUMVA	−64.580	142.62	$\alpha_1 = 2.4 \times 10^{-8}$, $\alpha_2 = 7.3 \times 10^{-8}$, $\sigma^2_1 = 0.0114$, $\sigma^2_2 = 0.0119$, $\theta_1 = 3.68$, $\theta_2 = 5.20$

in `geiger`. This method has been applied to paleontological data (Anderson et al. 2013; Benson and Choiniere 2013; Soul and Benson 2017; Ruta et al. 2019). A further example, trait MEDUSA, can be applied to multivariate data and is implemented using maximum likelihood in the package `motmot` (Thomas and Freckleton 2012; Puttick et al. 2020), and has also been applied to datasets of fossil taxa (Brocklehurst and Brink 2017; Button et al. 2017; Ruta et al. 2019). Both of these packages can also be used to fit and compare simpler phylogenetic models. Other methods can be used to infer rates for multivariate data, rather than to a single trait (Adams 2014). Diniz-Filho et al. (2015) use an approach based on phylogenetic eigenvectors to detect non-stationarity (i.e. significant shifts) in evolutionary rates in the skulls of theropod dinosaurs. Levy processes can be used to model jumps in evolutionary rate along branches in a phylogeny (Landis and Schraiber 2017), although this has not yet been implemented for nonultrametric trees.

9.4 Other New Approaches

There are a range of other phylogenetic models and rate estimation approaches that have been, or could be, applied to trees of fossil taxa. Often they can be used to model trait change in groups of lineages, characters, or periods of time, without having to specify the groupings in advance. This can be particularly useful for exploratory analyses of trait change, rather than for hypothesis testing as in our previous example analyses. At the time of writing, many of these methods are relatively new and have not yet been applied in a broad range of different scenarios. This means both that there are exciting opportunities for new discoveries and that in some cases the behavior of the methods across different areas of parameter space is not yet well characterized and results should be interpreted cautiously. Below is a subset of new approaches that have been applied to fossil data, or where they have not yet been applied to paleontological phylogenies, they have a software implementation that allows for it.

(a) The Fokker-Planck-Kolmogorov (FPK) model (Boucher et al. 2018; Blomberg et al. 2020) is a flexible modeling framework that can be used to estimate the shape of a macroevolutionary landscape that has one or more optima, using a parameter called the "evolutionary potential." This framework is distinct from the earlier multi-peak Ornstein-Uhlenbeck models where switches between macroevolutionary regimes are modeled; instead, it fits a model of trait change in a single macroevolutionary regime, but that regime can have multiple optima, and be bounded or unbounded.

(b) Mitov et al. (2019) introduced an approach for fitting mixed Gaussian phylogenetic models, which infers regimes that can be any combination of mode or rate for each model parameter across the phylogeny (for implementation see Mitov et al. 2019).

(c) Parins-Fukuchi (2020) introduces a modeling framework for change in disparity of continuous traits partitioned into integrated suites.

(d) The matching competition model (Drury et al. 2016, modified for application to nonultrametric trees in Manceau et al. 2017) is designed to model the effect of biotic interactions on trait evolution and is implemented in RPanda (Morlon et al. 2016).

(e) As a more explicit alternative to convergence inferred via multi-peak Ornstein-Uhlenbeck models, Stayton (2015) presents phylogenetic metrics designed for hypothesis tests of convergence. Speed and Arbuckle (2017) provide a review of other recent approaches to quantifying convergence using phylogeny.

(f) Relative rates of morphological evolution and ancestral character states can be estimated using the threshold model (Revell 2014, applied to paleontological data in Sallan et al. 2018).

10 Conclusion

As models for the "Tree of Life," phylogenetic trees are science's best attempt to represent a very complex process – the diversification of life through time – in a tractable way. Phylogenies and comparative methods allow us to answer questions about how evolution works that could not otherwise be answered, but they are obviously a simplification of real evolutionary processes. There is a large suite of PCMs available for paleontologists to apply in order to understand their group of interest, and most of these are now easily implemented using R and other software. These methods are very powerful tools for inference, but are not always intuitive. For this reason, they can seem intimidating, and misunderstandings are common. We hope that we have provided a practical entry point for those looking to begin using PCMs in their own work, while emphasizing the importance of careful interpretation of results and a good understanding of the relationship between the quality of the data you put in and the reliability of the information you get out. Phylogenies are themselves hypotheses, and uncertainty in the ages of fossils and relationships between the taxa they represent, combined with uncertainties in model fit, should always be considered when interpreting results and should always be clearly represented in published work. Caveats aside, this is an exciting time to be a paleontologist, as recent advances in computational and mathematical tools are allowing us to address some of evolutionary biology's biggest questions in a rigorous, quantitative way.

Supporting Materials

Supplemental data, full scripts for analyses, and trees are available at: github .com/daveyfwright/PCMsForPaleontologists.

References

Ackerly D. 2009. Conservatism and diversification of plant functional traits: Evolutionary rates versus phylogenetic signal. *Proc. Natl. Acad. Sci.* 106: 19699–19706.

Adams D. C. 2014. Quantifying and comparing phylogenetic evolutionary rates for shape and other high-dimensional phenotypic data. *Syst. Biol.* 63: 166–177.

Anderson P. S. L., Friedman M., Ruta M. 2013. Late to the table: Diversification of tetrapod mandibular biomechanics lagged behind the evolution of terrestriality. *Integr. Comp. Biol.* 53: 197–208.

Ausich W. I., Wright D. F., Cole S. R., Sevastopulo G. D. 2020. Homology of posterior interray plates in crinoids: A review and new perspectives from phylogenetics, the fossil record and development. *Palaeontology.* 63: 525–545.

Bapst D. W. 2012. Paleotree: An R package for paleontological and phylogenetic analyses of evolution. *Methods Ecol. Evol.* 3: 803–807.

Bapst D. W. 2013a. A stochastic rate-calibrated method for time-scaling phylogenies of fossil taxa. *Methods Ecol. Evol.* 4: 724–733.

Bapst D. W. 2013b. When can clades be potentially resolved with morphology? *PLoS One.* 8: e62312.

Bapst D. W. 2014a. Assessing the effect of time-scaling methods on phylogeny-based analyses in the fossil record. *Paleobiology.* 40: 331–351.

Bapst D. W. 2014b. Preparing palaeontological datasets for phylogenetic comparative methods. In: Garamszegi L. Z., editor. *Modern phylogenetic comparative methods and their application in evolutionary biology.* Berlin, Heidelberg: Springer-Verlag. pp. 515–544.

Bapst D. W., Hopkins M. J. 2017. Comparing cal3 and other a posteriori time-scaling approaches in a case study with the pterocephaliid trilobites. *Paleobiology.* 43: 49–67.

Barido-Sottani J., Pett W., O'Reilly J. E., Warnock R. C. M. 2019. FossilSim: An R package for simulating fossil occurrence data under mechanistic models of preservation and recovery. *Methods Ecol. Evol.* 10: 835–840.

Barido-Sottani J., Saupe E., Smiley T. M., Soul L. C., Wright A. M., Warnock R. C. M. 2020. Seven rules for simulations in paleobiology. *Paleobiology.* 46(4): 435–444.

Barido-Sottani J., Tiel N. van, Hopkins M. J., Wright D. F., Stadler T., Warnock R. C. M. 2020. Ignoring fossil age uncertainty leads to inaccurate topology and divergence times in time calibrated tree inference. *Frontiers in Ecology and Evolution*, 8: 183

Baum D. A., Smith S. D. 2013. *Tree thinking: An introduction to phylogenetic biology.* Greenwood Village, CO: Roberts..

Benson R. B. J., Choiniere J. N. 2013. Rates of dinosaur limb evolution provide evidence for exceptional radiation in Mesozoic birds. *Proc. R. Soc. B Biol. Sci.* 280: 20131780.

Blomberg S. P., Garland T., Ives A. R. 2003. Testing for phylogenetic signal in comparative data: Behavioral traits are more labile. *Evolution.* 57: 717–745.

Blomberg S. P., Rathnayake S. I., Moreau C. M. 2020. Beyond Brownian motion and the Ornstein-Uhlenbeck process: Stochastic diffusion models for the evolution of quantitative characters. *Am. Nat.* 195: 145–165.

Blomberg S. P., Lefevre J. G., Wells J. A., Waterhouse M. 2012. Independent contrasts and PGLS regression estimators are equivalent. *Syst. Biol.* 61: 382–391.

Boettiger C., Coop G., Ralph P. 2012. Is your phylogeny informative? Measuring the power of comparative methods. *Evolution.* 66: 2240–2251.

Boucher F. C., Démery V., Conti E., Harmon L. J., Uyeda J. 2018. A general model for estimating macroevolutionary landscapes. *Syst. Biol.* 67: 304–319.

Brocklehurst N., Brink K. S. 2017. Selection towards larger body size in both herbivorous and carnivorous synapsids during the Carboniferous. *Facets.* 2: 68–84.

Butler M. A., King A. A. 2004. Phylogenetic comparative analysis: A modeling approach for adaptive evolution. *Am. Nat.* 164: 683–695.

Button D. J., Barrett P. M., Rayfield E. J. 2017. Craniodental functional evolution in sauropodomorph dinosaurs. *Paleobiology.* 43: 435–462.

Clarke J. T., Lloyd G. T., Friedman M. 2016. Little evidence for enhanced phenotypic evolution in early teleosts relative to their living fossil sister group. *Proc. Natl. Acad. Sci.* 113: 11531–11536.

Close R. A., Friedman M., Lloyd G. T., Benson R. B. J. 2015. Evidence for a mid-Jurassic adaptive radiation in mammals. *Curr. Biol.* 25: 2137–2142.

Cole S. R., Wright D. F., Ausich W. I. 2019. Phylogenetic community paleo-ecology of one of the earliest complex crinoid faunas (Brechin Lagerstätte, Ordovician). *Palaeogeogr. Palaeoclimatol. Palaeoecol.* 521: 82–98.

Cooper N., Thomas G. H., FitzJohn R. G. 2016. Shedding light on the "dark side" of phylogenetic comparative methods. *Methods Ecol. Evol.* 7: 693–699.

Darwin C. R. 1859. *On the origin of species by means of natural selection.* London: John Murray.

Diniz-Filho J. A. F., Alves D. M. C. C., Villalobos F., Sakamoto M., Brusatte S. L., Bini L. M. 2015. Phylogenetic eigenvectors and nonstationarity in the evolution of theropod dinosaur skulls. *J. Evol. Biol.* 28: 1410–1416.

Drury J., Clavel J., Manceau M., Morlon H. 2016. Estimating the effect of competition on trait evolution using maximum likelihood inference. *Syst. Biol.* 65: 700–710.

Eastman J. M., Alfaro M. E., Joyce P., Hipp A. L., Harmon L. J. 2011. A novel comparative method for identifying shifts in the rate of character evolution on trees. *Evolution.* 65: 3578–3589.

Erwin D. H. 2007. Disparity: Morphological pattern and developmental context. *Palaeontology.* 50: 57–73.

Felsenstein J. 1985. Phylogenies and the comparative method. *Am. Nat.* 125: 1–15.

Finarelli J. A., Flynn J. J. 2006. Ancestral state reconstruction of body size in the Caniformia (Carnivora, Mammalia): The effects of incorporating data from the fossil record. *Syst. Biol.* 55: 301–313.

Foote M. 1996. On the probability of ancestors in the fossil record. *Paleobiology.* 22: 141–151.

Freckleton R. P. 2009. The seven deadly sins of comparative analysis. *J. Evol. Biol.* 22: 1367–1375.

Garamszegi L. Z. 2014. *Modern phylogenetic comparative methods and their application in evolutionary biology.* Berlin, Heidelberg: Springer-Verlag.

Garland T., Ives A. R. 2000. Using the past to predict the present: Confidence intervals for regression equations in phylogenetic comparative methods. *Am. Nat.* 155: 346–364.

Gascuel O., Steel M. 2014. Predicting the ancestral character changes in a tree is typically easier than predicting the root state. *Syst. Biol.* 63: 421–435.

Gavryushkina A., Welch D., Stadler T., Drummond A. 2014. Bayesian inference of sampled ancestor trees for epidemiology and fossil calibration. *PLoS Comput. Biol.* 10: e1003919.

Gavryushkina A., Heath T. A., Ksepka D. T., Stadler T., Welch D., Drummond A. J. 2017. Bayesian total-evidence dating reveals the recent crown radiation of penguins. *Syst. Biol.* 66: 57–73.

Gearty W., Payne J. L. 2020. Physiological constraints on body size distributions in Crocodyliformes. *Evolution.* 74: 245–255.

Halliday T. J. D., Goswami A. 2016. The impact of phylogenetic dating method on interpreting trait evolution: A case study of Cretaceous-Palaeogene eutherian body-size evolution. *Biol. Lett.* 12: 6–12.

Hansen T. F. 1997. Stabilising selection and the comparative analysis of adaptation. *Evolution.* 51: 1342–1351.

Hansen T. F., Martins E. P. 1996. Translating between microevolutionary process and macroevolutionary patterns: The correlation structure of inter-specific data. *Evolution*. 50: 1404–1417.

Harmon, Luke. 2019. "Phylogenetic Comparative Methods: Learning from Trees." EcoEvoRxiv. May 20. doi:10.32942/osf.io/e3xnr.

Harmon L. J., Weir J. T., Brock C. D., Glor R. E., Challenger W. 2008. GEIGER: Investigating evolutionary radiations. *Bioinformatics*. 24: 129–131.

Harmon L. J., Losos J. B., Davies T. J., Gillespie R. G., Gittleman J. L., Jennings B. W., Kozak K. H., McPeek M. A., Moreno-Roark F., Near T. J., Purvis A., Ricklefs R. E., Schluter D., Schulte II J. A., Seehausen O., Sidlauskas B. L., Torres-Carvajal O., Weir J. T., Mooers A. Ø. 2010. Early bursts of body size and shape evolution are rare in comparative data. *Evolution*. 64: 2385–2396.

Harrison L. B., Larsson H. C. E. 2015. Among-character rate variation distributions in phylogenetic analysis of discrete morphological characters. *Syst. Biol.* 64: 307–324.

Harvey P. H., Read A. F., Nee S. 1995. Further remarks on the role of phylogeny in comparative ecology. *J. Ecol.* 83: 733.

Heath T. A., Huelsenbeck J. P., Stadler T. 2014. The fossilized birth–death process for coherent calibration of divergence-time estimates. *Proc. Natl. Acad. Sci.* 111: E2957–E2966.

Hedman M. M. 2010. Constraints on clade ages from fossil outgroups. *Paleobiology*. 36: 16–31.

Ho L. S. T., Ané C. 2014a. A linear-time algorithm for Gaussian and non-Gaussian trait evolution models. *Syst. Biol.* 63: 397–408.

Ho L. S. T., Ané C. 2014b. Intrinsic inference difficulties for trait evolution with Ornstein-Uhlenbeck models. *Methods Ecol. Evol.* 5: 1133–1146.

Hopkins M. J., Smith A. B. 2015. Dynamic evolutionary change in post-Paleozoic echinoids and the importance of scale when interpreting changes in rates of evolution. *Proc. Natl. Acad. Sci. U.S.A.* 112: 3758–3763.

Hunt G. 2012. Measuring rates of phenotypic evolution and the inseparability of tempo and mode. *Paleobiology*. 38: 351–373.

Hunt G. 2013. Testing the link between phenotypic evolution and speciation: An integrated palaeontological and phylogenetic analysis. *Methods Ecol. Evol.* 4: 714–723.

Hunt G., Carrano M. T. 2010. Models and methods for analyzing phenotypic evolution in lineages and clades. *Paleontol. Soc. Pap.* 16: 245–269.

Hunt G., Slater G. 2016. Integrating paleontological and phylogenetic approaches to macroevolution. *Annu. Rev. Ecol. Evol. Syst.* 47: 189–213.

Jeremy M. Beaulieu and Brian O'Meara (2020). OUwie: Analysis of Evolutionary Rates in an OU Framework. R package version 2.5. https://CRAN.R-project.org/package=OUwie.

Kammer T. W. 2008. Paedomorphosis as an adaptive response in pinnulate cladid crinoids from the Burlington limestone (Mississippian, Oseadean) of the Mississippi Valley. In: Webster, G. D., Maples, C. D., editors. *Echinoderm paleobiology.* Bloomington, IN: University of Indiana Press. pp. 177–195.

Lande R. 1976. Natural selection and random genetic drift in phenotypic evolution. *Evolution.* 30: 314.

Landis M. J. 2017. Biogeographic dating of speciation times using paleogeographically informed processes. *Syst. Biol.* 64: 307–324.

Landis M., Schraiber J. G. 2017. Pulsed evolution shaped modern vertebrate diversity. *Proc. Natl. Acad. Sci. U.S.A.* 114: 13224–13229.

Lewis P. O. 2001. A likelihood approach to estimating phylogeny from discrete morphological character data. *Syst. Biol.* 50: 913–925.

Lloyd G. T. 2016. Estimating morphological diversity and tempo with discrete character-taxon matrices: Implementation, challenges, progress, and future directions. *Biol. J. Linn. Soc.* 118: 131–151.

Lloyd G. T., Wang S. C., Brusatte S. L. 2012. Identifying heterogeneity in rates of morphological evolution: Discrete character change in the evolution of lungfish (Sarcopterygii; Dipnoi). *Evolution.* 66: 330–348.

Maddison D. R., Maddison W. P. 2020. *MacClade* 4. http://macclade.org/macclade.html.

Manceau M., Lambert A., Morlon H. 2017. A unifying comparative phylogenetic framework including traits coevolving across interacting lineages. *Syst. Biol.* 66: 551–568.

Matzke N. J., Wright A. 2016. Inferring node dates from tip dates in fossil Canidae: The importance of tree priors. *Biol. Lett.* 12: 1–4.

Mitov V., Bartoszek K., Stadler T. 2019. Automatic generation of evolutionary hypotheses using mixed Gaussian phylogenetic models. *Proc. Natl. Acad. Sci.* 116: 16921–16926.

Morlon H., Lewitus E., Condamine F. L., Manceau M., Clavel J., Drury J. 2016. RPANDA: An R package for macroevolutionary analyses on phylogenetic trees. *Methods Ecol. Evol.* 7: 589–597.

Nielsen R. 2002. Mapping mutations on phylogenies. *Syst. Biol.* 51: 729–739.

Nunn C. L. 2011. *The comparative approach in evolutionary anthropology and biology.* Chicago: University of Chicago Press.

Nunn C. L., Barton R. A. 2001. Comparative methods for studying primate adaptation and allometry. *Evol. Anthropol.* 10: 81–98.

O'Meara B. C., Ané C., Sanderson M. J., Wainwright P. C. 2006. Testing for different rates of continuous trait evolution using likelihood. *Evolution.* 60: 922–933.

O'Reilly J. E., Puttick M. N., Parry L., Tanner A. R., Tarver J. E., Fleming J., Pisani D., Donoghue P. C. J. 2016. Bayesian methods outperform parsimony but at the expense of precision in the estimation of phylogeny from discrete morphological data. *Biol. Lett.* 12: 20160081.

Paradis E., Claude J., Strimmer, K. 2004. APE: Analyses of phylogenetics and evolution in R language. *Bioinformatics.* 20: 289–290.

Parins-Fukuchi C. 2020. Detecting mosaic patterns in macroevolutionary disparity. *Am. Nat.* 195: 129–144.

Pennell M. W., Fitzjohn R. G., Cornwell W. K., Harmon L. J. 2015. Model adequacy and the macroevolution of angiosperm functional traits. *Am. Nat.* 186: E33–E50.

Pennell M. W., Eastman J. M., Slater G. J., Brown J. W., Uyeda J. C., FitzJohn R. G., Alfaro M. E., Harmon L. J. 2014. Geiger V2.0: An expanded suite of methods for fitting macroevolutionary models to phylogenetic trees. *Bioinformatics.* 30: 2216–2218.

Pinheiro J., Bates D., DebRoy S., Sarkar D. 2019. nlme: Linear and nonlinear mixed effects models. *R package version* 3: 1–140.

Polly P. D. 2019. Spatial processes and evolutionary models: A critical review. *Palaeontology.* 62: 175–195.

Price S. A. 2019. State-dependent diversification of traits. http://treethinkers .org/tutorials/state-dependent-diversification-of-traits.

Price S. A., Friedman S. T., Wainwright P. C. 2015. How predation shaped fish: The impact of fin spines on body form evolution across teleosts. *Proc. R. Soc. B Biol. Sci.* 282. https://doi.org/10.1098/rspb.2015.1428.

Puttick M. N. 2016. Partially incorrect fossil data augment analyses of discrete trait evolution in living species. *Biol. Lett.* 12: 20160392.

Puttick M. N., Ingram T., Clarke M., Thomas G. H. 2020. MOTMOT: Models of trait macroevolution on trees (an update). *Methods Ecol. Evol.* 11: 464–471.

Puttick M. N., O'Reilly J. E., Pisani D., Donoghue P. C. J. 2019. Probabilistic methods outperform parsimony in the phylogenetic analysis of data simulated without a probabilistic model. *Palaeontology.* 62: 1–17.

Revell, L.J. (2010), Phylogenetic signal and linear regression on species data. Methods in Ecology and Evolution, 1: 319–329.

Revell L. J. 2012. phytools: An R package for phylogenetic comparative biology (and other things). *Methods Ecol. Evol.* 3: 217–223.

Revell L. J. 2013. Two new graphical methods for mapping trait evolution on phylogenies. *Methods Ecol. Evol.* 4: 754–759.

Revell L. J. 2014. Ancestral character estimation under the threshold model from quantitative genetics. *Evolution.* 68: 743–759.

Revell L. J., Schliep K., Valderrama E., Richardson J. E. 2018. Graphs in phylogenetic comparative analysis: Anscombe's quartet revisited. *Methods Ecol. Evol.* 9: 2145–2154.

Rohlf F. J. 2006. A comment on phylogenetic correction. *Evolution.* 60: 1509.

Ruta M., Krieger J., Angielczyk K. D., Wills M. A. 2019. The evolution of the tetrapod humerus: Morphometrics, disparity, and evolutionary rates. *Earth Environ. Sci. Trans. R. Soc. Edinburgh.* 109: 351–369.

Sallan L., Friedman M., Sansom R. S., Bird C. M., Sansom I. J. 2018. The nearshore cradle of early vertebrate diversification. *Science.* 464: 460–464.

Silvestro D., Kostikova A., Litsios G., Pearman P. B., Salamin N. 2015. Measurement errors should always be incorporated in phylogenetic comparative analysis. *Methods Ecol. Evol.* 6: 340–346.

Simpson G. G. 1944. *Tempo and mode in evolution.* New York: Columbia University Press.

Slater G. J. 2013. Phylogenetic evidence for a shift in the mode of mammalian body size evolution at the Cretaceous-Palaeogene boundary. *Methods Ecol. Evol.* 4: 734–744.

Slater G. J. 2014. Correction to "Phylogenetic evidence for a shift in the mode of mammalian body size evolution at the Cretaceous-Palaeogene boundary," and a note on fitting macroevolutionary models to comparative paleontological data sets. *Methods Ecol. Evol.* 5: 714–718.

Slater G. J., Pennell M. W. 2014. Robust regression and posterior predictive simulation increase power to detect early bursts of trait evolution. *Syst. Biol.* 63: 293–308.

Slater G. J., Harmon L. J., Alfaro M. E. 2012. Integrating fossils with molecular phylogenies improves inference of trait evolution. *Evolution.* 66: 3931–3944.

Soul L. C., Benson R. B. J. 2017. Developmental mechanisms of macroevolutionary change in the tetrapod axis: A case study of Sauropterygia. *Evolution.* 71: 1164–1177.

Soul L. C., Friedman M. 2015. Taxonomy and phylogeny can yield comparable results in comparative palaeontological analyses. *Syst. Biol.* 64: 608–620.

Soul L. C., Friedman M. 2017. Bias in phylogenetic measurements of extinction and a case study of end-Permian tetrapods. *Palaeontology.* 60: 169–185.

Speed M. P., Arbuckle K. 2017. Quantification provides a conceptual basis for convergent evolution. *Biol. Rev.* 92: 815–829.

Stadler T. 2010. Sampling-through-time in birth-death trees. *J. Theor. Biol.* 267: 396–404.

Stadler T., Gavryushkina A., Warnock R. C. M., Drummond A. J., Heath T. A. 2018. The fossilized birth-death model for the analysis of stratigraphic range data under different speciation modes. *J. Theor. Biol.* 447: 41–55.

Stayton C. T. 2015. The definition, recognition, and interpretation of convergent evolution, and two new measures for quantifying and assessing the significance of convergence. *Evolution.* 69: 2140–2153.

Thomas G. H., Freckleton R. P. 2012. MOTMOT: Models of trait macroevolution on trees. *Methods Ecol. Evol.* 3: 145–151.

Uyeda J. C., Zenil-Ferguson R., Pennell M. W. 2018. Rethinking phylogenetic comparative methods. *Syst. Biol.* 67: 1091–1109.

Voje K. L., Starrfelt J., Liow L. H. 2018. Model adequacy and micro-evolutionary explanations for stasis in the fossil record. *Am. Nat.* 191: 509–523.

Wagner P. J. 2012. Modelling rate distributions using character compatibility: implications for morphological evolution among fossil invertebrates. *Biol. Lett.* 8: 143–146.

Wagner P. J., Marcot J. D. 2010. Probabilistic phylogenetic inference in the fossil record: current and future applications. *Paleontol. Soc. Pap.* 16: 189–211.

Wagner, P.J. and Marcot, J.D., 2013. Modelling distributions of fossil sampling rates over time, space and taxa: assessment and implications for macroevolutionary studies. *Methods in Ecology and Evolution*, 4(8), pp.703–713.

Warnock R. C. M., Wright A. M. 2020. Understanding the tripartite approach to Bayesian divergence time estimation. *EcoEvoRxiv.* https://doi.org/10.32942 /osf.io/4vazh.

Wesley-Hunt G. D. 2005. The morphological diversification of carnivores in North America. *Paleobiology.* 31: 35–55.

Westoby M., Leishman M., Lord J. 2016. Further remarks on phylogenetic correction. *J. Ecol.* 83: 727–729.

Wiley E. O., Lieberman B. S. 2011. *Phylogenetics: Theory and practice of phylogenetic systematics.* New York: John Wiley & Sons.

Wright A. M. 2019. A systematist's guide to estimating Bayesian phylogenies from morphological data. *Insect Syst. Divers.* 3: 2.

Wright A. M., Hillis D. M. 2014. Bayesian analysis using a simple likelihood model outperforms parsimony for estimation of phylogeny from discrete morphological data. *PLoS One.* 9: e109210.

Wright A. M., Lloyd G. T., Hillis D. M. 2016. Modeling character change heterogeneity in phylogenetic analyses of morphology through the use of priors. *Syst. Biol.* 65: 602–611.

Wright A. M., Wagner P. J., Wright D. F. 2020. Testing character evolution models in phylogenetic paleobiology: A case study with Cambrian echinoderms. *EcoEvoRxiv.* https://doi.org/10.32942/osf.io/ykzg5.

Wright D. F. 2015. Fossils, homology, and phylogenetic paleo-ontogeny: A reassessment of primary posterior plate homologies among fossil and living crinoids with insights from developmental biology. *Paleobiology.* 41: 570–591.

Wright D. F. 2017a. Bayesian estimation of fossil phylogenies and the evolution of early to middle Paleozoic crinoids (Echinodermata). *J. Paleontol.* 91: 799–814.

Wright D. F. 2017b. Phenotypic innovation and adaptive constraints in the evolutionary radiation of palaeozoic crinoids. *Sci. Rep.* 7: 1–10.

Wright D. F., Toom U. 2017. New crinoids from the Baltic region (Estonia): Fossil tip-dating phylogenetics constrains the origin and Ordovician–Silurian diversification of the Flexibilia (Echinodermata). *Palaeontology.* 60: 893–910.

Acknowledgments

This Element stems from materials originally developed for a 2019 Paleontological Society sponsored workshop titled "Quantitative Methods in Phylogenetic Paleobiology." We thank Vera Korasidis and Melanie J. Hopkins for providing feedback on an early version of this Element, as well as Will Gearty and David Bapst for thorough reviews that improved the manuscript's quality and clarity. LCS acknowledges support from the Smithsonian National Museum of Natural History Deep Time Initiative. DFW acknowledges support from the Gerstner Scholars Fellowship and the Gerstner Family Foundation, the Lerner-Gray Fund for Marine Research, and the Richard Gilder Graduate School, American Museum of Natural History, as well as a Norman Newell Early Career Grant from the Paleontological Society.

Cambridge Elements ☰

Elements of Paleontology

Editor-in-Chief
Colin D. Sumrall
University of Tennessee

About the Series
The Elements of Paleontology series is a publishing collaboration between the Paleontological Society and Cambridge University Press. The series covers the full spectrum of topics in paleontology and paleobiology, and related topics in the Earth and life sciences of interest to students and researchers of paleontology.

The Paleontological Society is an international nonprofit organization devoted exclusively to the science of paleontology: invertebrate and vertebrate paleontology, micropaleontology, and paleobotany. The Society's mission is to advance the study of the fossil record through scientific research, education, and advocacy. Its vision is to be a leading global advocate for understanding life's history and evolution. The Society has several membership categories, including regular, amateur/avocational, student, and retired. Members, representing some 40 countries, include professional paleontologists, academicians, science editors, Earth science teachers, museum specialists, undergraduate and graduate students, postdoctoral scholars, and amateur/avocational paleontologists.

Paleontological
S O C I E T Y

Cambridge Elements $\equiv$

Elements of Paleontology